Industrial Water Use and Reuse Workshop 2013

San Antonio, Texas, USA
30 April – 2 May 2013

ISBN: 978-1-5108-1603-9

Printed from e-media with permission by:

Curran Associates, Inc.
57 Morehouse Lane
Red Hook, NY 12571

Some format issues inherent in the e-media version may also appear in this print version.

Printed by Curran Associates, Inc. (2015)

AIChE
120 Wall Street, FL 23
New York, NY 10005-4020

Phone: (800) 242-4363
Fax: (203) 775-5177

www.aiche.org

Additional copies of this publication are available from:

Curran Associates, Inc.
57 Morehouse Lane
Red Hook, NY 12571 USA
Phone: 845-758-0400
Fax: 845-758-2634
Email: curran@proceedings.com
Web: www.proceedings.com

TABLE OF CONTENTS

AIChE International Society for Water Solutions

Industrial Water Use and Reuse Workshop

April 30-May 2, 2013

Ralph E. Exton
Chief Marketing Officer
GE Power & Water, Water & Process Technologies

International Society for Water Solutions

1

Industrial Water Use and Reuse Workshop

Tuesday, April 30

- Evening Reception

Wednesday, May 1

- Opening Remarks
- Keynote Address
- Technical Sessions
  - Regulatory Climate
  - Industry Perspectives
  - Technology and Applications
- Networking Reception

Thursday, May 2

- Opening Remarks
- Technical Sessions
  - Practical Implementation
  - Expert Panel
- Wrap Up

International Society for Water Solutions

2

Industrial Water Use and Reuse Workshop

Session Chairs
- Jim Goodrich, EPA
- Walker Garrison, Valero
- Dave Mason, GE Power & Water
- Loraine Huchler, MarTech Systems
- Tracy Young, Dow

Keynote Address
- Amanda Brock, Water Standard

Organizing Commitee
- Jeanette Brown, UConn
- Walker Garrison, Valero
- James Goodrich, EPA
- Loraine Huchler, MarTech Systems
- Annette Johnston, Abbott
- Wendy Reed, Chemstations
- Ari Selvakumar, EPA
- Laura Watson, GE Power & Water
- Tracy Young, Dow
- Janet Goodwin, EPA

International Society for Water Solutions

3

Industrial Water Use and Reuse Workshop

Experts / Presenters
- Amanda Brock, Water Standard
- Jon Freedman, GE Power & Water
- Julie Klumpyan, TRDWC
- James Brown, EPA
- Travis Capps, Valero
- Gena Leathers, DOW
- Carson Barry, New Page Corp
- Robin Jenkins, Dupont
- Jim Myers, Chevron
- Erik Hanson, GE Power & Water
- James Mihelcic, Univ. of S. Florida
- Larry Gurnari, Veolia
- Steve Gluck, Dow
- Joe Walters, West Basin MWD
- Al Pristera, Nalco
- Tim Eggert, GE Power & Water
- Jerry Levine, Water & Power Consultants
- Tara Lee Lanier, Marathon Petroleum
- Mikel Goldblatt, Ashland
- Sidd Dunn, Baker Petrolite

International Society for Water Solutions

4

Global water scarcity today – WRI Water Atlas

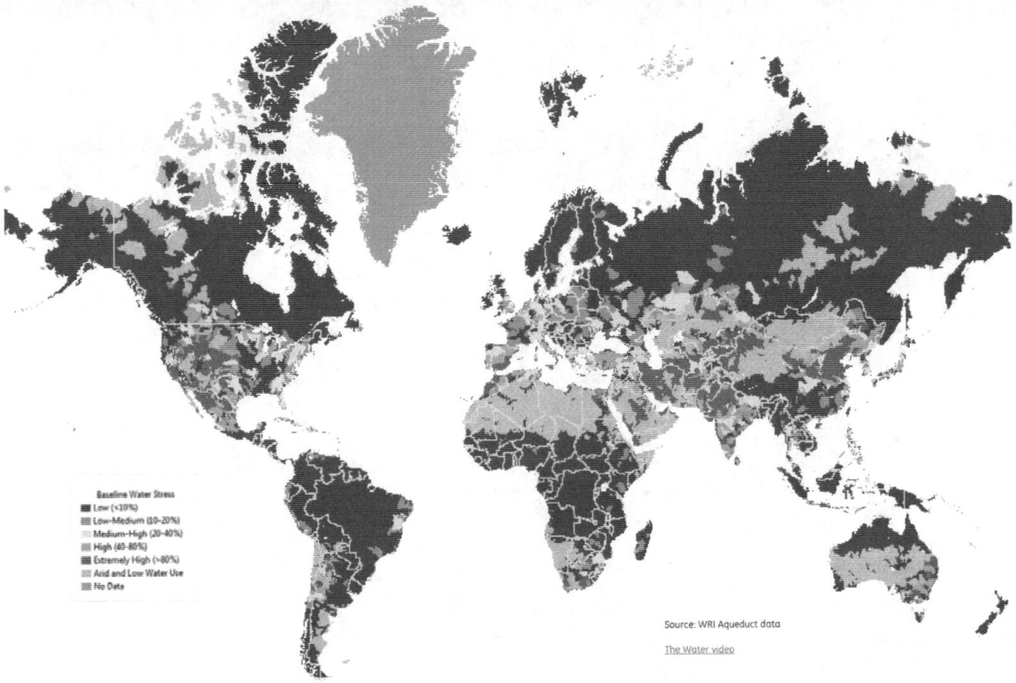

Source: WRI Aqueduct data

The Water video

Industrial Water Use and Reuse Workshop

A very local perspective...

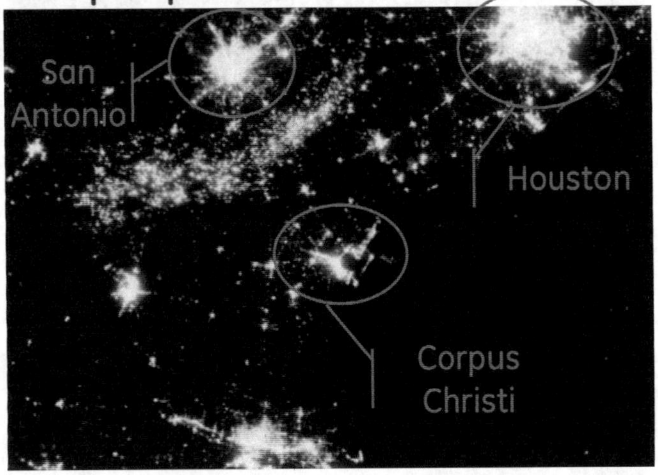

2007 2012

6

Water Reuse: The Public Perception

imagination at work

Survey Methodology

- StrategyOne conducted an online survey among a total of 3,000 adults, 18 years or older, 1,000 each in the US, Singapore, and China. Interviewing took place from June 1–11, 2012.
- The margin of sampling error at the 95% confidence level is as follows:
  - US (n=1,000): MOE = +/- 3.1%
  - Singapore (n=1,000): MOE = +/- 3.1%
  - China (n=1,000): MOE = +/- 3.1%
- The data set for each country is nationally representative of the population of that country.

imagination at work

Americans Strongly Support Reusing Water to Combat Water Scarcity

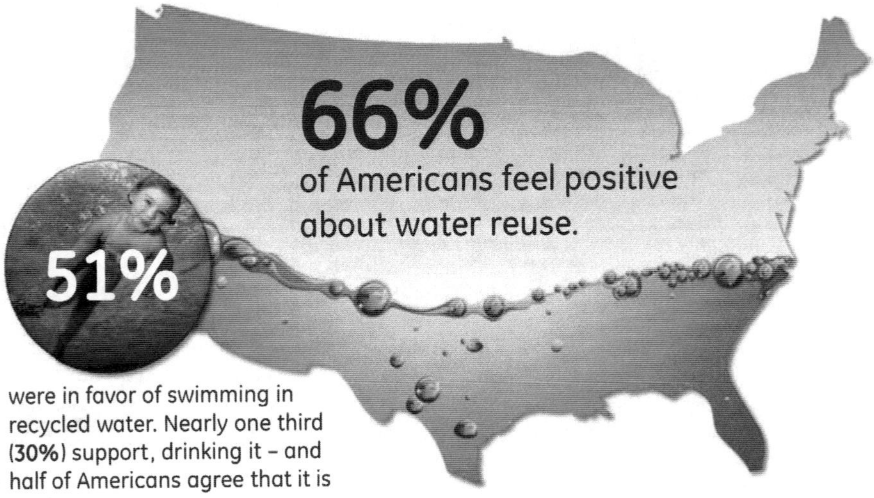

were in favor of swimming in recycled water. Nearly one third (**30%**) support, drinking it – and half of Americans agree that it is drinkable (**51%**).

Americans' Understanding of Water Lifecycle and Solutions Falls behind Water Challenged Countries...

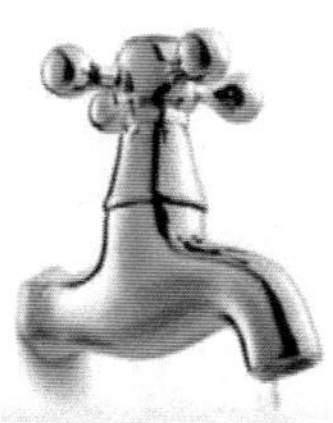

31% of Americans don't know where their water comes from, compared to only 14% of those in China and 15% of those in Singapore.

42% of **Americans don't know how their water is purified**, compared to only **17%** of those in China and **19%** of those in Singapore.

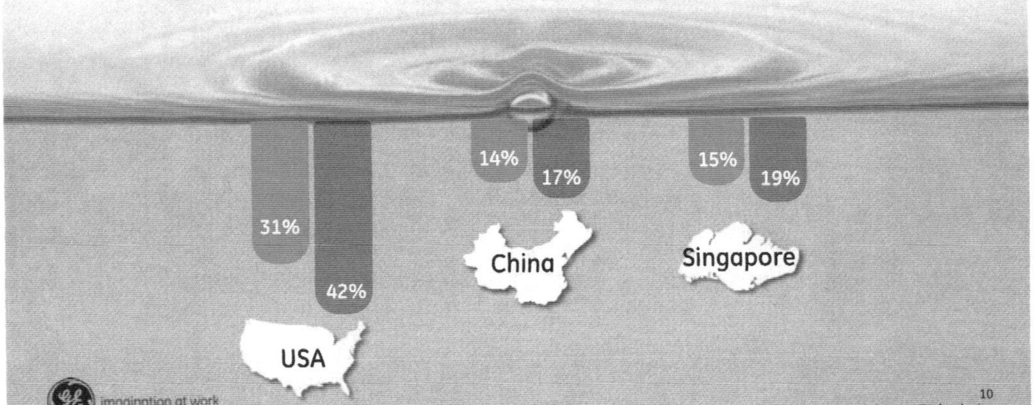

Concerned about Sufficient Water Supplies, Americans Call on National Government to Step Up and Lead on Water Resources to Protect Economic Competitiveness

84% of Americans believe the protection of water resources should be a national priority.

70% of Americans feel the US would have a competitive advantage over other countries if we increased the amount of water we reuse in a safe and efficient manner.

imagination at work

11

Water & Process Technologies

Americans understand intimate connection between Energy and Water, expect Power Industry Players to demonstrate Water Stewardship

74% of Americans are aware that you need water to create energy.

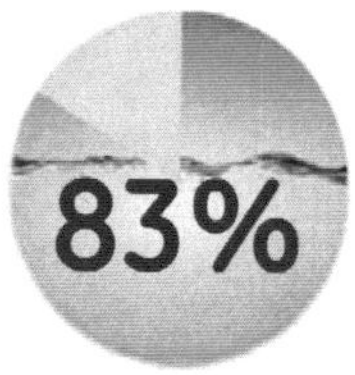

Americans believe smart water management can help the US more efficiently create and use energy.

imagination at work

12

Water & Process Technologies

Demand on water resources

Agriculture dominates global withdrawals ...

... But, industry dominates in advanced economies

Global Water Withdrawals
Percentage (%)

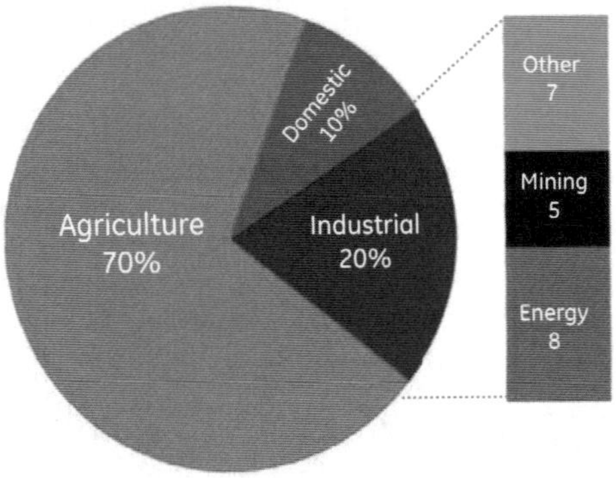

Source: Pacific Institute, The World's Water Vol 7
http://www.worldwater.org/data.html

Advanced Economies Water Withdrawals*
Percentage (%)

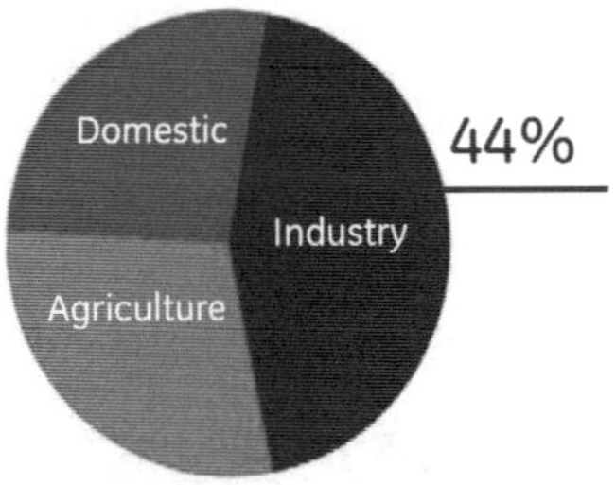

* IMF definition for advanced economies

Developing economies are moving in the same direction

imagination at work

14

Energy is water intensive

U.S. Domestic Per Capita Water Use		U.S. Power Sector's Per Capita Water Use
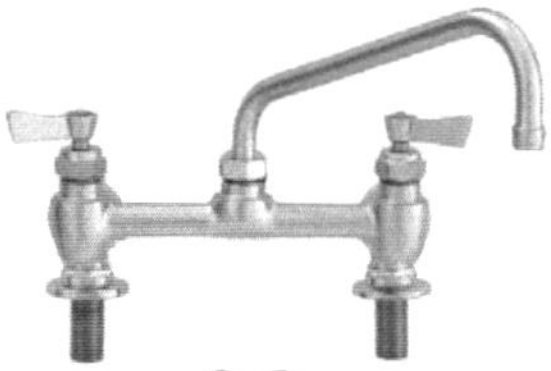	vs.	
98 gals /cap/day		560 gals /cap/day

Sources: EIA, US Census, USGS

Power sector's water requirements are 6x greater than current domestic usage

imagination at work

15

International Society for Water Solutions

The International Society for Water Solutions (ISWS) provides a forum for communication and networking among engineers and scientists who share interests in water research and water technology development.

Membership benefits include:

1. A professional network allowing you to share ideas and experiences with your colleagues from around the world.
2. A platform for individuals to present their work and peer review projects relevant to industry and academia.
3. Discounts on ISWS water conferences and workshops.
4. News that is relevant and contains "what you need to know" to stay current.
5. Webinars, available live and on-demand, featuring cutting-edge water technologies covered by leading experts.
6. Access to career and employment services.

Leadership and Sponsors

Danny Reible, Bettie Margaret Smith Chair of Environmental Health Engineering, The University of Texas at Austin, Chair ISWS

Ralph Exton, Chief Marketing Officer, GE Power & Water, Co-Chair ISWS

2013 ISWS Workshop Series

Industrial Water Use and Reuse
April 30th- May 2nd, 2013
Grand Hyatt, San Antonio, Texas

Municipal Water Reuse
September 23rd-25th, 2013
Hyatt Regency Long Beach, Long Beach, California

Water Energy Nexus
October 20th-24th, 2013
Hilton Alexandria Old Town, Alexandria, Virginia

Thanks to a generous grant from the United Engineering Foundation, the American Institute of Chemical Engineers, in conjunction with AIME and ASCE, will present a series of 3 workshops in 2013. Learn more about them here. The United Engineering Foundation does not endorse results or recommendations from these workshops.

NSF Energy-Water Nexus Workshop

June 10-11, 2013

National Science Foundation Headquarters
Alexandria, Virginia

Keynote Address

International Society for Water Solutions

Amanda Brock
CEO, Water Standard

Named one of the top 10 women in energy by the Houston Chronicle, Amanda has spent her career as a leader and pioneer in energy and water related businesses.

In 2007, she helped develop WATER STANDARD, and in March 2008, Ms. Brock raised one of the largest equity commitments for a start up in the water industry to globalize its business.

WATER STANDARD is now recognized as an industry leader in water treatment for water-based enhanced oil recovery and produced water treatment/reuse.

In 2012, Global Water Intelligence described WATER STANDARD as the company "doing the most to bridge the gap between the oil and the water industries."

21

Thank you

imagination at work

Back up slides

imagination at work

Water & Process Technologies

Aligning the best people with the most advanced technology, products and services to solve any water or process challenge our customers may have

Quick Facts
• Headquartered in Trevose, PA, USA • 8,000 employees globally • 50,000 customers in 130 countries • 50 global manufacturing sites

imagination at work

24
Water & Process Technologies

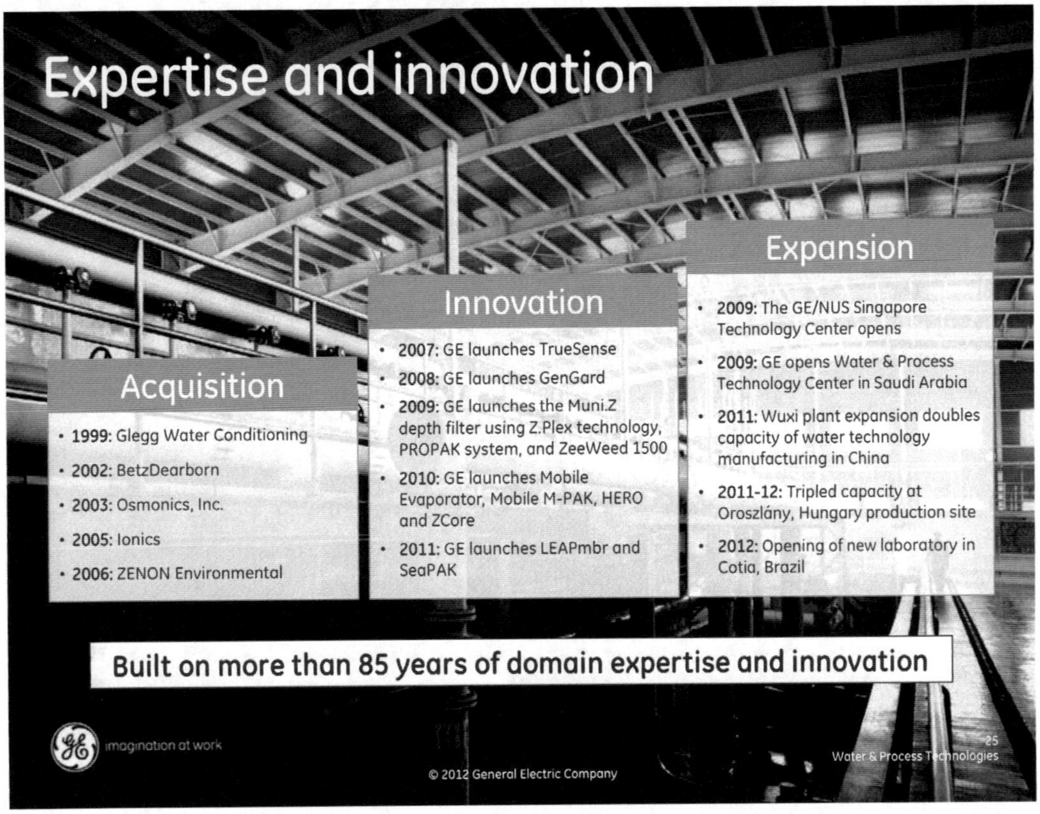

Global leadership position

Our leadership in equipment solutions:

- Advanced ultrafiltration, membrane bioreactor, reverse osmosis membranes and membrane chemistries
- Mobile fleet and water outsourcing capabilities
- Tough-to-treat applications, such as unconventional fuels and mining
- Packaged water treatment equipment
- Analytical instruments for measuring water quality

Our leadership in chemical and knowledge management solutions:

- Cooling and boiler water technologies that enable customers to protect their assets
- Chemical treatment for ethylene, styrene and elastomer production facilities
- Refinery treatment solutions focused on tough-to-treat crudes
- Remote monitoring and diagnostic solutions

imagination at work

26
Water & Process Technologies

Broad product and solutions portfolio

Chemical & Monitoring Solutions

- Cooling Chemistries
- Boiler Chemistries
- Wastewater Chemistries
- Fuel Oil Treatment
- Hydrocarbon Process Chemistries
- Industrial Process Chemistries
- Knowledge Management & Monitoring Solutions

Engineered Systems

- Ultrafiltration/Membrane Bioreactor
- Mobile Water Solutions
- Water Outsourcing
- Thermal/Zero Liquid Discharge
- Reverse Osmosis/Electrolytic Systems
- Filters & Membranes
- ABMet for selenium removal
- Analytical Instruments

imagination at work

WATER STANDARD

May 1, 2013

Impact of America's Oil Bonanza

- The shale-gas revolution in America has been as "sudden and startling as a supertanker performing a handbrake turn."[1]
- The International Energy Agency (IEA) predicts that the U.S. will become the world's largest oil producer by 2020, outstripping Saudi Arabia and Russia.
- The North American hydrocarbon bonanza offers big benefits, but also some pitfalls. The economic pluses are obvious: cheap gas yields cheap electricity, which boosts American industry, especially power-hungry sectors such as aluminum, steel and glass.
- The negatives are not so obvious, but very real. The impact on water supply, reuse and disposal is key.

1) Economist

Shale Boom Powering an Industrial Revolution

- There is an industrial revival going on spurred by the U.S. shale boom.
- New life for the Petrochemical, Steel and ancillary manufacturing industries.
  - Methanex closed its last U.S. chemical plant in 1999; now spending $500M to move a methanol plant from Chile to Louisiana
  - Williams spending $400M to expand an ethylene plant
  - CF Industries spending $2.1B to expand its fertilizer capacity
  - $80B spend anticipated in chemical, fertilizer, steel, aluminum, tire and plastics plants
- U.S. steel chairman, John Surma stated, "Development of our Nations recoverable oil and natural gas resources has the potential to be the once-in-a-lifetime economic engine that coal was 200 years ago."
- "With the price of gas likely to remain 50 to 70% cheaper in the U.S. than Europe and Japan, that will translate into significantly lower costs for electricity generation, fuel to power industrial plants and for feed stock used across many industrial processes."[1]
- The shale boom is changing how we think and act.

Water availability will be a key factor in realizing this anticipated industrial growth

Source: Justin Rose, BGG

The Industrial Water Challenge

- We live in an era of increasing water scarcity; it's not going away.
- Climate change is real and expensive!
- There is a critical focus on balancing supply, demand and managing competing interests of agriculture, domestic and industrial users.
- Industrial users will always be lower on the allocation and perception scale in any period of restricted use.
- It's all about the food, water, energy and industrial nexus.
- The new challenge is how to manage a growing industrial base in the U.S. and to streamline decision making taking into consideration all commodities.
- We have to understand supply and demand and make informed decisions accordingly.

Understanding the Basic's

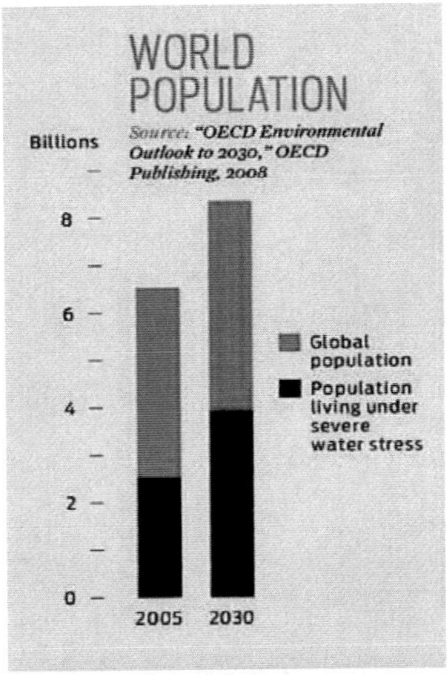

- Water scarcity is a growing problem and is exacerbated by macro-economic drivers.
- The world's population will increase by an estimated 2.5 billion people by 2050.
- By 2025, over half the world's population will live in water-stressed or water-scarce regions.
- Today 50% of the global population lives in coastal areas with this number increasing annually through migration toward the coast.
- Rising standards of living is resulting in increased water consumption.
- As a result of climate change, droughts are extending in duration due to changes in the hydrological cycle.
- 2 billion tons of contaminated water is discharged daily across the world. More people now die from contaminated and polluted water than from all forms of violence including wars.

Water is a non-substitutable commodity.
Supply/Demand imbalances dictate the opportunities and challenges.

Source: IBRD/World Bank, Sick Water (UNDP), UNESCO
OECD Environmental Outlook to 2030, IEEE Spectrum Special Report – June 2010

Population Growth: Driving Demand for Water

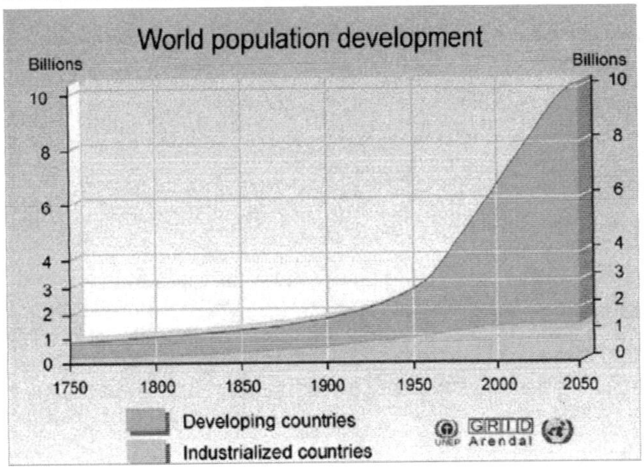

- The world's population has tripled in the last a hundred years yet there is no more fresh water on Earth now than there was 2000 years ago.
- Today's population of 6.7 billion is already placing unsustainable demands on water resources, currently using over 2.1 trillion cubic meters of water per year—the equivalent of two times the volume of Lake Superior (the largest lake in the world).

Source(s): The Sustainable Scale Project (graphics)
UN Urban Population, Development and the Environment, 2007

Population Growth: Driving Demand for Industrial Growth

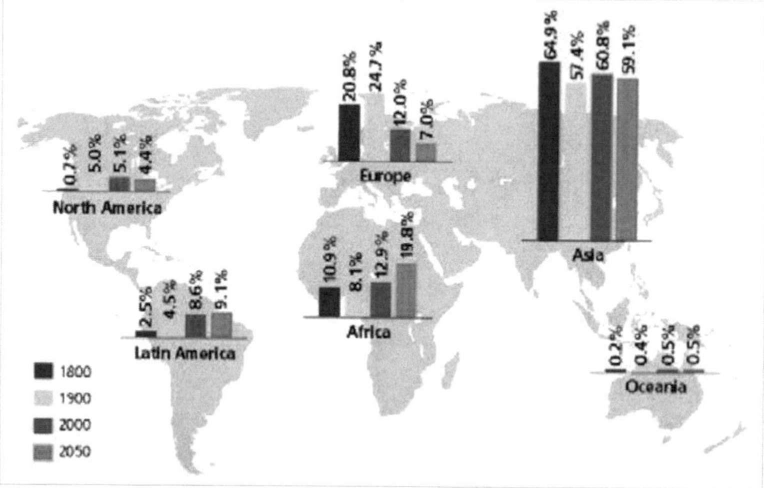

- By 2050, the population is expected to increase to 9.2 billion.
- The additional 2.5 billion persons—*the equivalent to two more Chinas*—will be absorbed mostly by the less developed regions, where infrastructure and water resources are unlikely to be able to handle such a large increase in population.

Source(s): The Sustainable Scale Project (graphics)
UN Urban Population, Development and the Environment, 2007

Resource Allocation: Unequal Distribution Stress Factors

Global water supply versus global population by region

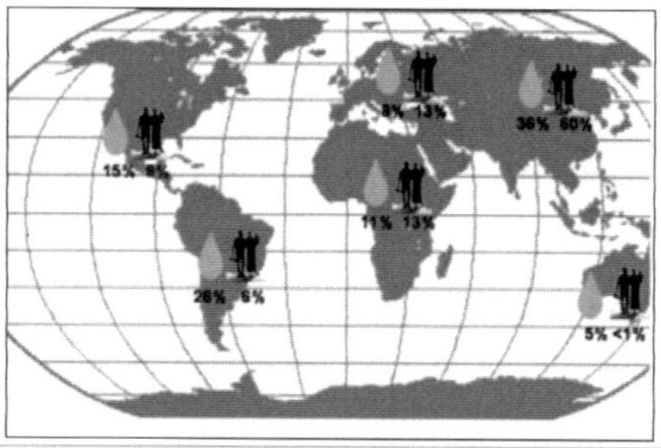

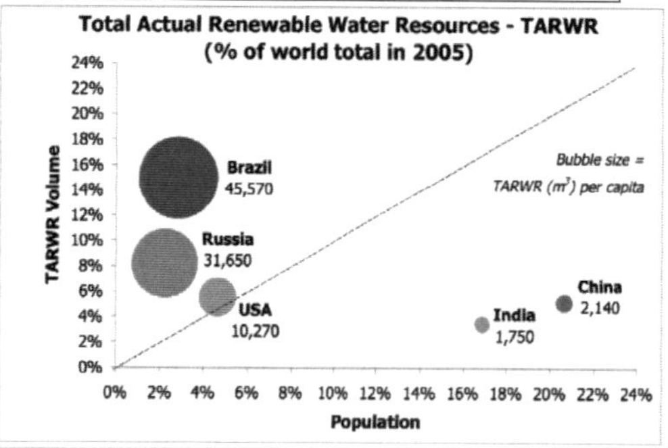

- Fresh water resources are often not located near areas of population concentration.
- Two-thirds of the world's population live in areas that receive only one-quarter of the world's annual rainfall.
- Over half of the world's water resources are located in only six countries.
- Russia enjoys a ratio of 9% of the world's water supply for only around 2% of the world's population.
- China faces severe water shortages with over 22% of world population and only 8% of world fresh water resources.

Source(s): UN Educational, Scientific and Cultural Organization
FAO-AQUASTAT 2005, Fusion Analysis

Rising Standards of Living

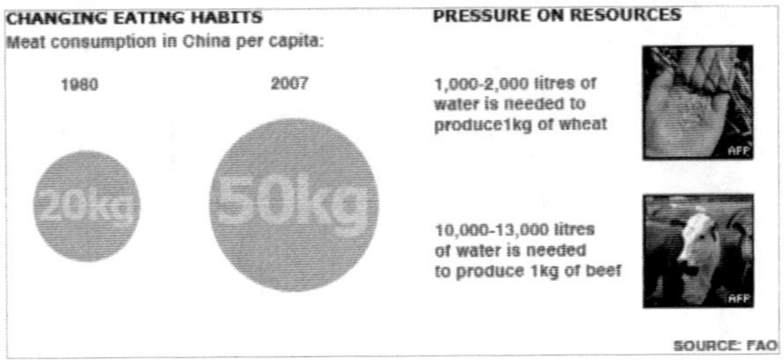

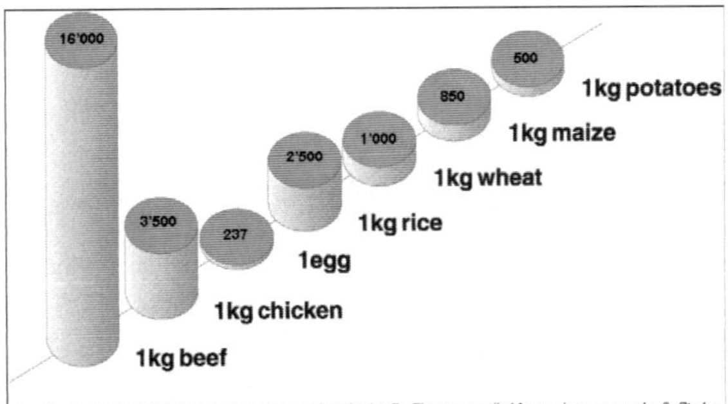

Quantity of water (in liters) necessary to produce various foodstuffs. Figures compiled from various sources by C. Studer.

- By 2050, food demand is expected to increase by 70%-90% due to population growth and increasing incomes.
- Crop water consumption is expected to grow by the same order of magnitude.
- Higher income levels result in more varied and protein-rich diets, increasing the amount of water it takes to satisfy food requirements.
- Consumption of meat, a far more water-intensive good than grain, in China has grown from 20kg per capita to 50kg per capita in about 25 years.
- As standards of living increase, demand for industrial products increase; power; and consumables.

Source(s): United Nations Water Development BBC

Population Concentration on Coastlines

- Populations worldwide are concentrated along or near coasts occupying only 10% of the earth's land. Industry follows people.

Ten Largest Cities

1) Tokyo	***Coastal***	***4) Shanghai***	***Coastal***	7) Sáo Paulo	Inland
2) Mexico City	Inland	***5) Lagos***	***Coastal***	***8) New York, USA***	***Coastal***
3) Mumbai	***Coastal***	***6) Los Angeles***	***Coastal***	***9) Calcutta***	***Coastal***
				10) Buenos Aires	***Coastal***

- This concentration pollutes and depletes coastal and ocean resources and places pressure on fresh water resources, leading to water crises for many coastal communities.
- Coastal population will grow to around 7 billion globally by 2040, further compounding problems of fresh water access.

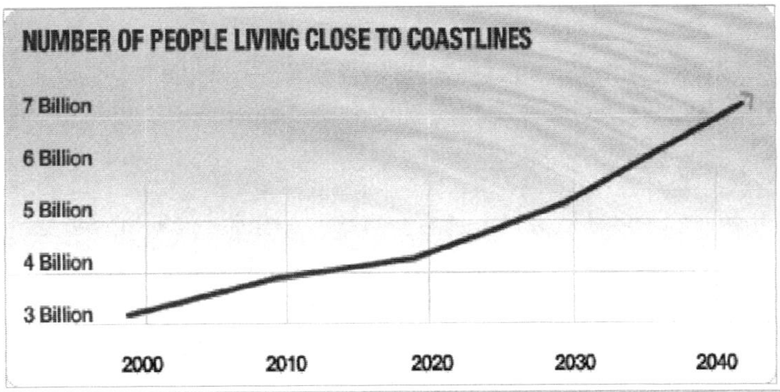

Source(s): Solutions for a Water-Short World, Population Reports US Department of Commerce Nahkeel

Climate Change

With record droughts and the changes in water patterns, the continuing effect of climate change has an immediate impact on the availability and delivery patterns of fresh water.

Murray-Darling Basin - Australia
August 2010

Amazon River -Brazil

September 2006

Lake Mead - Nevada

April 2007

Kouris Reservoir - Cyprus

March 2008

Climate Change Impacts

- Rising costs
- Calculation of "Water Footprint" for all users
- Regulatory caps for water use
- Regulatory restrictions for specific industrial activities
- Conflicts with local communities and regional and political implications
- Impact on site locations and permitting for industry
- Operational disruptions and associated financial loss

Water and its availability and quality will be the main pressures on societies, the environment and industry under climate change.

Source: IPCC Technical Paper, "*Climate Change and Water*", June 2008; Ceres Water Scarcity and Climate Change: "*Growing Risks for Businesses and Investors*", April 2009; *Research and Markets Report 2007 Desalination Markets 2007 A Global Industry Forecast*

Water Scarcity has a Real Impact

- In the 2011 drought, Texas suffered an estimated $7.62 billion in agricultural losses.
- The Texas 2012 State Water Plan prepared by the Texas Water Development Board, concluded, if drought conditions recur and water management strategies are not implemented, Texas could suffer annual economic losses of $11.9 Billion growing to $115.7 Billion by 2060.
- Regional planning groups identified 562 projects and estimated it will take $53 Billion to implement recommended water management strategies and projects.
- Texas is now also looking at a growing manufacturing base, expecting 12.8% more drilling activity in the Eagleford and 15.1% in the Permian for 2013, and needs to grow its power generation capacity.

Patterns of Water Usage: Increasing Competition

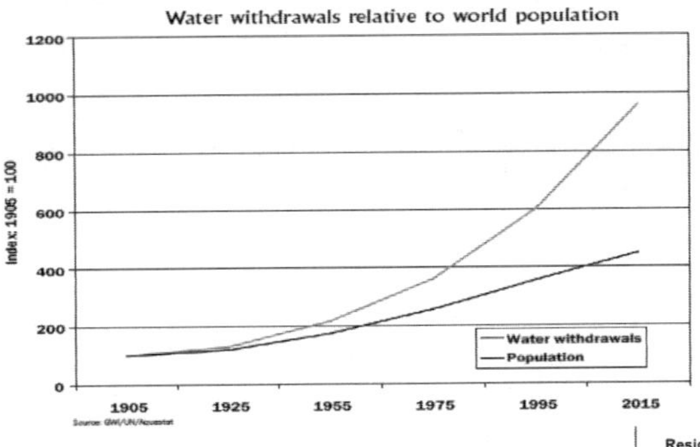

- The same forces of population growth and increasing incomes will also raise demand for manufactured goods and energy consumption.
- Industrial water needs are expected to increase 50% in the next 10 years.

- Competition for water resources between industry and agriculture will only intensify in the future with a very real threat of chronic food shortages and surge in prices due to water scarcity.

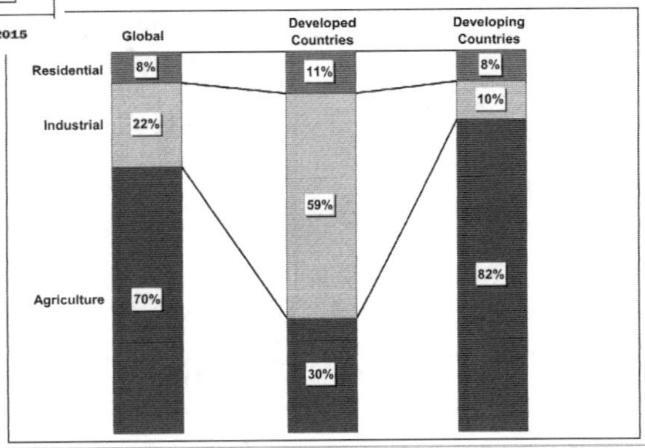

Allocation of Water Resources Globally

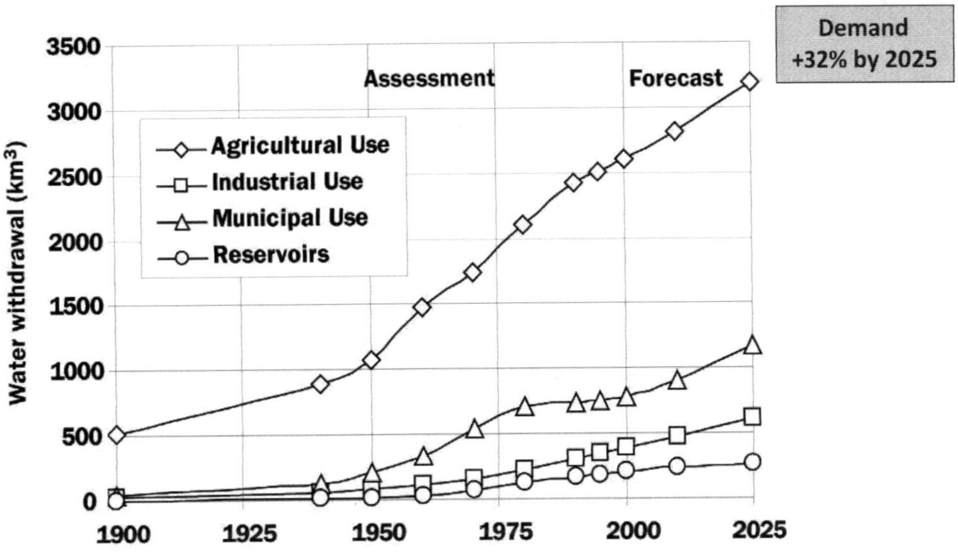

Global Demand: Where is the Growth?

Increase in Annual Water Demand 2005-2030

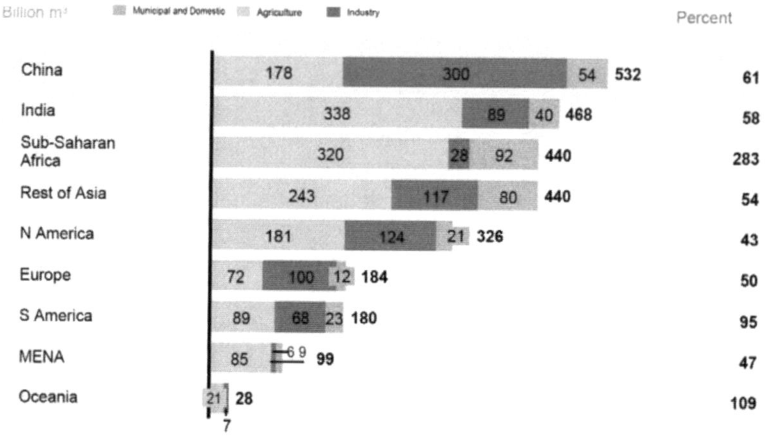

SOURCE: 2030 Water Resources Global Water Supply and Demand model; baseline agricultural production based on IFPRI IMPACT-WATER base case

How will the shale boom change these estimates?

Increasing U.S. Water Withdrawals

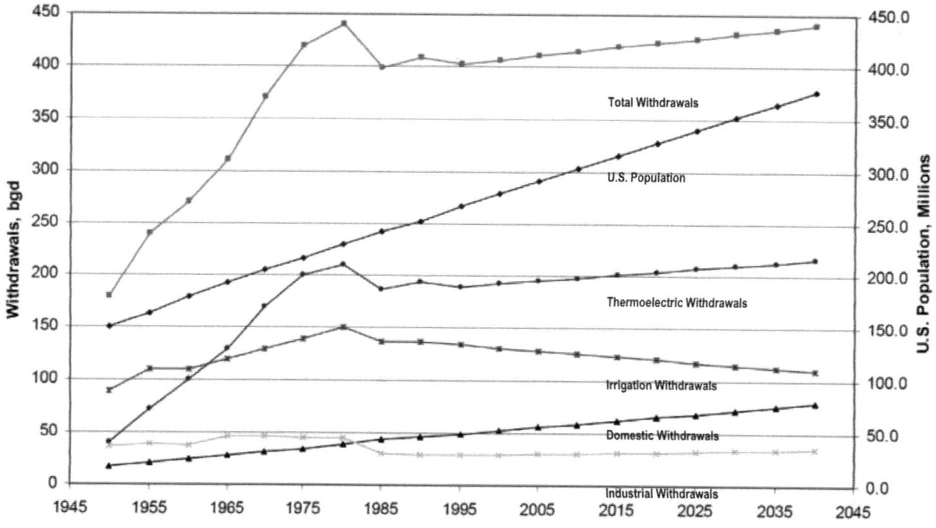

United States industrial demand had reduced as a result of outsourcing production to developing nations, but that is changing.

Source: Dziegielewski et al. 2003, Institute of Water Resources of the USACE U.S. Department of Interior

Sources of Water Demand in U.S.

Patterns in the U.S. are different from developing countries

- 1 m^3 of industrial water is 70x more valuable than 1 m^3 of irrigation
- To produce I kw takes 21 liters of water
- To produce 1 kg of steel takes 95 liters of water
- To produce 1 kg of paper takes 326 liters water
- Emergence of the concept of virtual water
- "Industry" is the easy target in an environment where there are competing interests

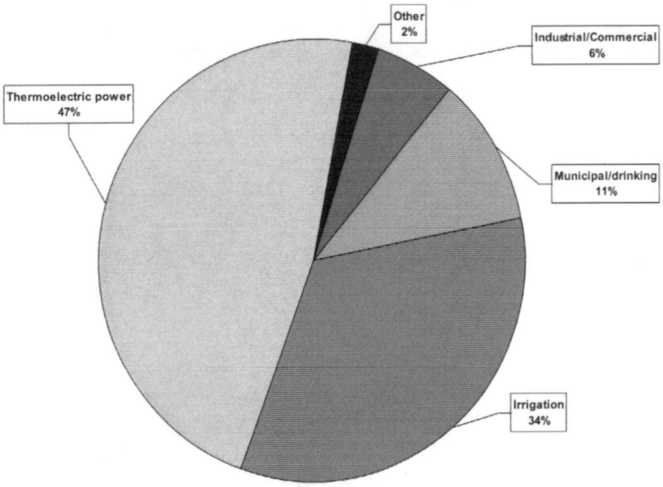

65% of water in the U.S. is currently surface water, 20% groundwater and 15% saline

Water Use for Energy Industry

Water volumes needed for energy to double by 2035!

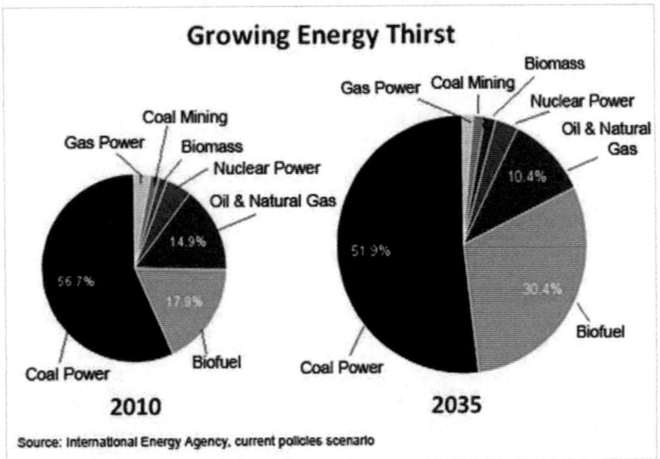

National Geographic

This did not account for the rapid growth of the shales

Today's Growth Environment

Macro Uncertainty Abounds

- World economies are fragile
  - Global deleveraging cycle will be a drag on growth everywhere
  - Europe is a mess with no resolution in sight
  - China is a black box – possibly headed for a "hard landing"
  - U.S. recovery is nascent, slow and hooked on fiscal stimulus
- Politicians not inspiring confidence
- Despite this, U.S. stock market is hitting new highs and debt yields are very low
- U.S. Oil & Gas Boom is providing the only bright spot and industry is growing

Source: Limerock

The Battle Lines are Being Drawn

- Who wins when water is scarce; not industry.
- Recently an example in Texas highlighted the potential risks:

  Dow Chemical made a priority call for water. The Texas Commission on Environmental Quality (TCEQ) determined that Dow could not divert the amount of water it needed, but allowed junior water rights holders to continue to use their appropriations. They were Municipal and Electric users and TCEQ argued public health, safety and welfare issues were at play.
- The implications are significant for industry, but in this case also agriculture; rice farmers.
- Appointment of water masters and water czars.
- Lack of effective Federal and State leadership.

The Market is Taking Note

- We have to better understand and assess the impact of water scarcity on future corporate growth and adopt effective water risk management practices.
- Ceres, an advocate for sustainability leadership, has focused on making corporate disclosure of water risk a priority for investors and also a mechanism for effective water management.
- Investors want to understand how well companies are managing water related risks and opportunities.
- Ceres has developed a first-of-its-kind tool called the Ceres Aqua Gauge which is designed, in part, to help investors analyze corporate water risk.
- Ceres recently stated that this standardized tool is endorsed by investors managing over $2 Trillion in assets and helps these investors assess the impact of water risks on the companies they are evaluating.

Industry will have to Adapt

Water shortages are not going away.

Additional water supply for increasing demand can only be obtained through four (4) basic sources:

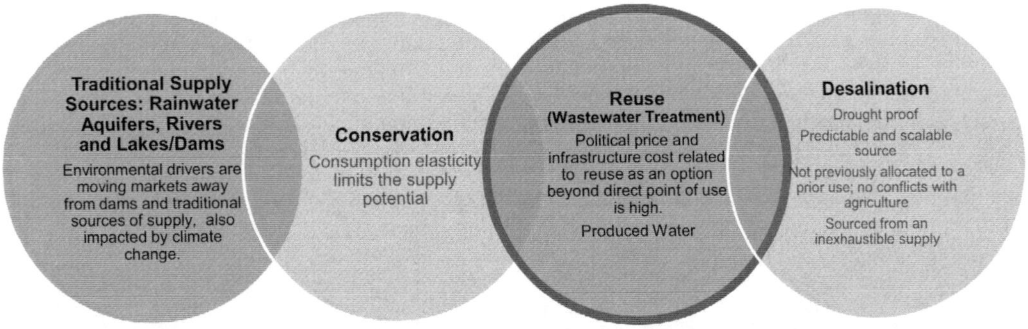

Water availability will drive decisions on where to locate new industries, but for existing infrastructure, all options have to be considered.

Focus on Efficiency and Reuse

- An estimated 20mgd of fresh water is pumped through U.S. industries today. That is going to increase significantly.
- The water withdrawn is pumped back into municipal wastewater treatment facilities or disposed into bodies of water after treatment.
- In a water short environment, rising regulatory compliance obligations and increasing cost, water reuse has to be considered.
- Need for water security and re-thinking wastewater as a source of potential profit and just smart water risk management.
  - Ecolab and Dow in Freeport addressed water use within Nalco's 3D Trasar Cooling Water Technology and yielded a savings of 1 billion gallons of water per annum and $4M per annum in OPEX.
- Increase in innovation and technology related to treatment and reuse.

Upstream Sector Analogy

- Competing interests for accessing water supplies and bad press have resulted in fundamental changes in the use of water in hydraulic fracturing.
  - EPA estimates it takes between 70 and 140 billion gallons to fracture 35,000 wells a year.
  - Dealing with Flowback is challenging and cost of treatment or disposal is significant.
  - Result is a focus on:
    - Reuse of flowback after treatment
    - Reinjection of flowback with minimal treatment
    - Moving away from water to use of propane gas, compressed air or CO_2 foam
  - Tremendous R&D and new technology advances to treat water.
- Produced Water Reuse
  - U.S. generates approximately 21B bbl per annum with Produced Water as the highest volume by product associated with production (98%).
  - Increased environmental regulation and requirement for treatment before discharge; movement to ZLD offshore.
  - Innovation in treatment methods for reinjection and Enhanced Oil Recovery.

Where does this Leave the Industrial Sector

- There is a lot of uncertainty and a lot of opportunity.
- Water use has to be looked at holistically and industrial users have to face the reality they will always be vulnerable to criticism.
- In the face of the lack of effective political leadership and regulatory certainty, the industry will have to be ahead of the curve, self-regulate and innovate.
- In an environment of dwindling water resources it is not only a supply issue, but also a demand side management issue.
- As the U.S. industrial base grows and competition for water as a limited resource grows, industry has to manage water risk.
- Effective Water Management = profit and success and drives long-term decisions.

WATER STANDARD

FOR MORE INFORMATION CONTACT

Amanda Brock, CEO
ambrock@waterstandard.com

Executive Office:
4265 San Felipe, Suite 620
Houston, Texas USA 77027
+1-713-400-4777
www.waterstandard.com

RIGHT FOR THE ENVIRONMENT | RIGHT FOR SAFETY | RIGHT FOR THE INDUSTRY

Disclaimer

This presentation was not peer reviewed or approved for publication. It is an informative presentation intended to educate the public on water planning and reuse trends. The contents of this presentation do not necessarily reflects the views and policies of the U.S. Environmental Protection Agency, nor does mention of trade names, commercial productions, corporations or publications constitute endorsement or recommendation for use.

Outline

- Water Footprint
- Continued Drought/Depleted Supplies
- Increasing Demand/Population Growth
- Energy Costs
- Corporate Stewardship
- Regulations and Guidelines
- Water Reuse Examples

"Water Footprint"

Beef	*1,799 gallons/lb*
Hamburger	*660*
Chicken	*468*
Wheat	*132*
Potato	*119*
Rice	*449*
Milk	*880 gallons/gallon*
Wine	*1008*
Beer	*689*

Natural gas	*0.15 gal/kWh*
Oil	*1.01*
Coal	*0.15*
Nuclear	*0.09*
Hydroelectric	*20.92*
Solar	*0.26*
Wind	*0*

http://waterfootprint.org or
http://environment.nationalgeographic.com/environment/freshwater/embedded-water/

Texas Drought

- As of April 13, 2013, 99% of TX is in some form of drought conditions, and the state's reservoirs are 66% full. Nearly 11% of the state is in exceptional drought, the worst stage.

- 2011 was the driest year ever for Texas, with an average of only 14.8 inches of rain. The only comparable drought occurred during the drought of record during the 1950s, but no single year during that drought was as dry as 2011

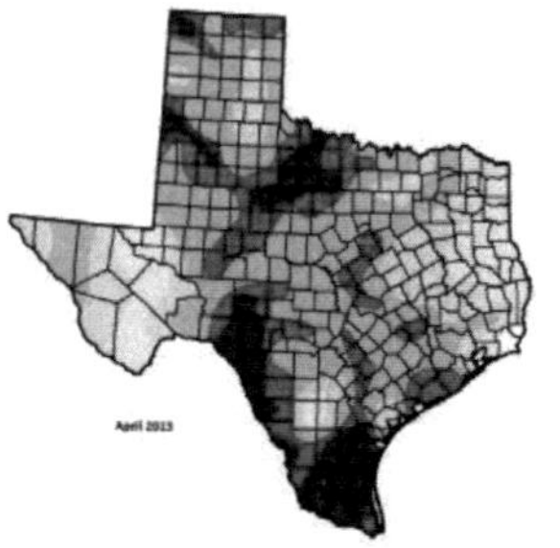

The latest NOAA outlook predicts the drought will "persist or intensify" in Texas in the coming months, and drought-free areas of the state are likely to see drought development.

http://droughtmonitor.unl.edu

Growth Projections in Texas

(~82% Increase from 2010 to 2060)

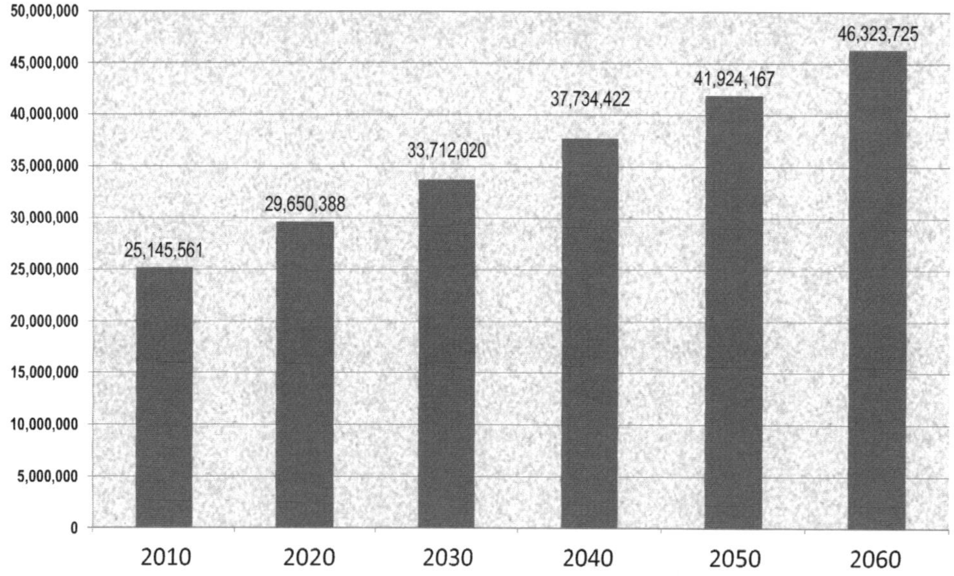

Future Water Demand in Texas

Water for Texas 2012 State Water Plan

Category	2010	2060	% Increase
Municipal	4,851,201	8,414,492	73%
Manufacturing	1,727,808	2,882,524	67%
Mining	296,230	292,294	-1%
Steam Electric	733,179	1,620,411	121%
Livestock	322,966	371,923	15%
Irrigation	10,079,215	8,370,554	-17%
Texas	18,010,599	21,952,198	22%

acre-feet/year

EPA WaterSense

Projected New Water Supplies to meet the Demands of 2060

Water for Texas 2012 State Water Plan

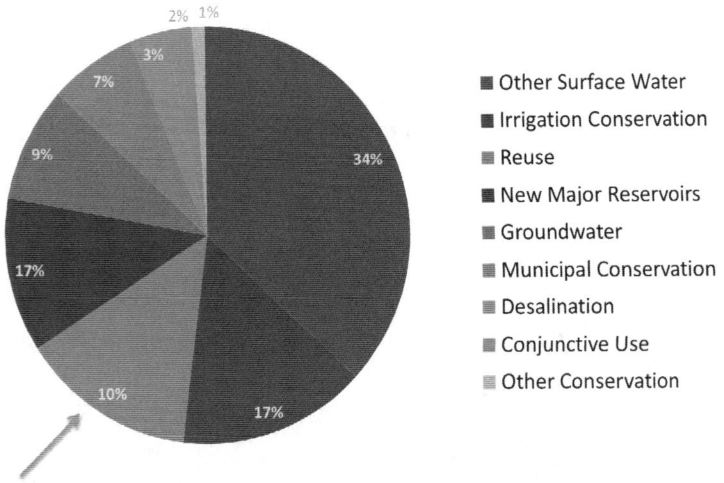

Water Reuse increases from 5% of total in 2010 to 10% of total in 2060 (~100,600 to ~915,000 acre-feet)

The Energy-Water Connection: Municipal water/sewer plant energy use

- U.S. annual total* = 75 billion kilowatt hours per year
- 3% to as much as 14% of total U.S. consumption of electricity
- Equal to entire residential electricity demand of California
- More than entire energy-intensive pulp/paper and petroleum sectors *combined*
- Public bill = Already $4B/yr. Increasing
- Residential water heating uses 104 Bkwh annually!

* 60,000 drinking water treatment plants + 15,000 sewage treatment plants

Sources: US EPA, Alliance to Save Energy, Pacific Institute & NRDC

Hidden Energy Costs in Water Use

- *Running a hot water faucet for 5 minutes consumes the same energy as a 60W light that's on for 14 hours!*

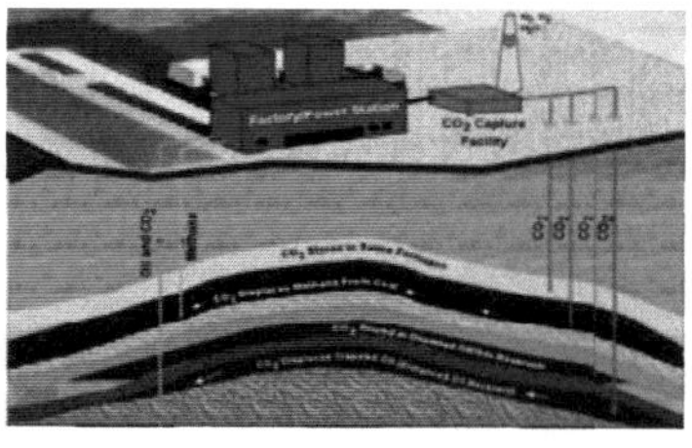

EPA WaterSense[SM]

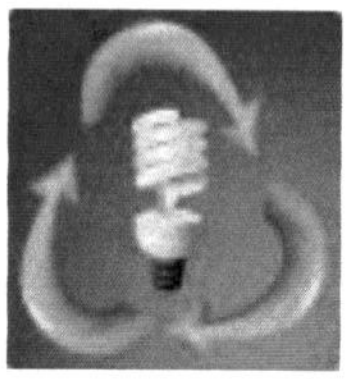

Corporate Social Responsibility

Some CSR Trends, Jan. 2012, Forbes Magazine Article

- **Globalization:** increasing accountability for supply chains.
- **Transparency:** Last year more than 5,500 companies around the world issued sustainability reports, up from about 800 a decade ago.
- **Employee Engagement Emerges:** morale was 55% better, business process were 43% more efficient, public image was 43% stronger, and employee loyalty was 38% better.
- **A Growing Population, 7 Billion:** Everything from energy efficient semiconductors to electric cars to water conserving plants will find new markets in 2012 and beyond.

Reuse Water

Reuse water is used as a conservation measure for potable water supplies.

Fire protection
Irrigation for Golf courses, Schools
Parks and athletic fields, nurseries, landscaping
Street sweeping
Agriculture, Irrigation, Livestock
Cooling towers
Power Plant Cooling
Scrubbers
Dust suppression at construction projects
Supplementing base flows in creeks and rivers
Industrial manufacturing processes
Potable water

Water Reuse Regulations and Guidelines

http://nepis.epa.gov/Adobe/PDF/P100FS7K.pdf

Water Reuse: Endless Possibilities

Aquifer Storage & Recovery

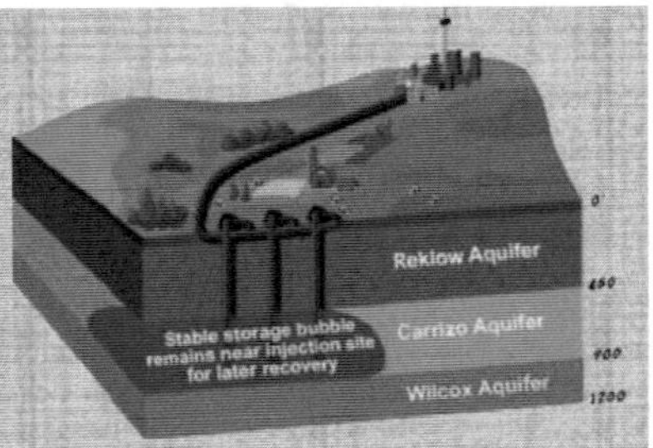

San Antonio, Texas: Twin Oaks facility

- *Operates 29 injection wells*
- *Stores water from Edwards aquifer in the Charrizo-Wilcox aquifer*
- *Around 30 billion gallons of water stored underground.*

El Paso, Texas: Fred Hervey water Reclamation Plant

- *Stores reclaimed water in the Hueco Bolson aquifer*
- *Over 500 million gallons stored underground*

Kerrville, Texas (upper Guadalupe River Authority)

- *Operates 2 injection wells*
- *Stores treated surface water*
- *Around 800 million gallons water stored underground*

El Paso, TX Kay Bailey Hutchison Desalination Plant

City of El Paso Water Utilities and Fort Bliss

- *Went into production in 2007*
- *Largest inland municipal desalination plant in the world*
- *Design capacity of approx. 27.5 million gallons per day (approx 30,800 acre-feet per year)*

Big Spring Water Cycle

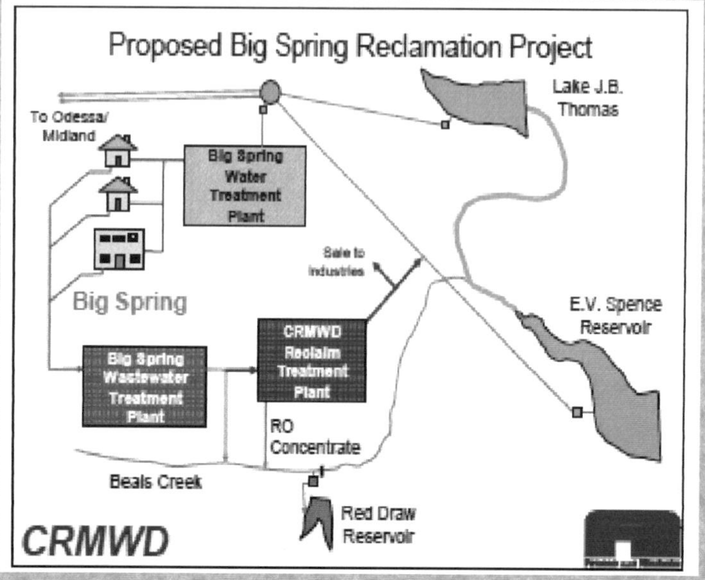

Construction has been completed on a $13 million water-reclamation facility. That's a fancy way of describing a treatment plant that will turn sewage wastewater into drinking water. It's in a testing phase now but is fully operational.

Water Reuse

Dos Rios WWTP and Sustainability

Sells every gallon of reclaimed water it produces to 65 big customers including the Alamo, and the River Walk.

98% of the sludge is recycled and sold through a commercial compost supplier. SAWS cut the cost of disposal by more than half.

The bio-solids digesters produce 1.5 Mft^3/day of natural gas. SAWS sells it for $20,000 a mo. They are the only municipal wastewater plant in the United States to sell its bio-gas commercially.

No technological breakthrough was required to turn Dos Rios from a "waste" facility into a "recycling" facility. The only thing that had to be overhauled was the thinking of the people.

70% of the cleaned water returns to the Medina River through a 45' high spillway. They are trying to figure out if they could put a small hydropower generator at this outfall.

Industrial Examples

2012 Water Reuse Guidelines; Chapter 3, section 3.5

Frito-Lay in Casa Grande, AZ Frito-Lay Process Water Recovery Treatment Plant, Casa Grande, Arizona. Set a goal of running a manufacturing facility almost entirely on renewable energy and reclaimed water and near zero waste.

Coca-Cola in GA Recovery and Reuse of Beverage Process Water: Installing recycle- and-reclaim loops in 12 of its water treatment systems in North America and Europe with goals of equipping up to 30 facilities. The loops save an average of 57 million gallons of water per system annually.

Industrial Examples

2012 Water Reuse Guidelines; Chapter 3, section 3.5

Intel: Improved the efficiency of the process used to create the ultra-pure water (UPW) required to clean silicon wafers during fabrication.

Previously, almost 2 gallons of water were needed to make 1 gallon of UPW. Today, Intel generates 1 gallon of UPW from between 1.25 and 1.5 gallons.

In 2010, Intel internally recycled approximately 2 billion gallons of water, equivalent to 25 percent of its total water withdrawals for the year.

look for
WaterSense
Meets EPA Criteria
Consumer Choices . . .
look for
WaterSense
Meets EPA Criteria
WaterSense
Meets EPA Criteria

WaterSense
Meets EPA Criteria
Do one thing!

International Society for Water Solutions

Regulations Driving Reuse

GE Power and Water

Water & Process Technologies

May 1, 2013

Jon Freedman

imagination at work

GE Water is a global leader

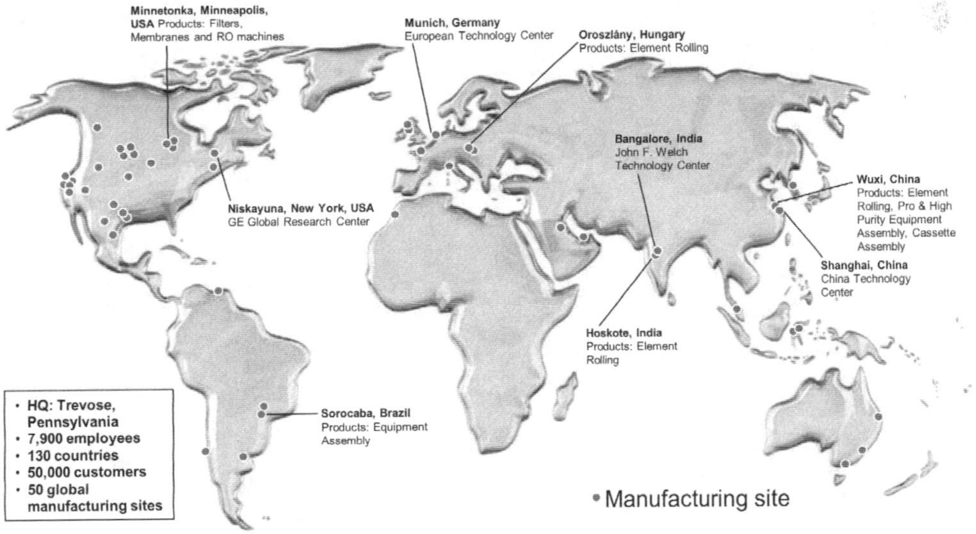

- **HQ: Trevose, Pennsylvania**
- **7,900 employees**
- **130 countries**
- **50,000 customers**
- **50 global manufacturing sites**

8,000 employees & 50,000 customers in 130 countries

imagination at work

2

And a broad portfolio of solutions

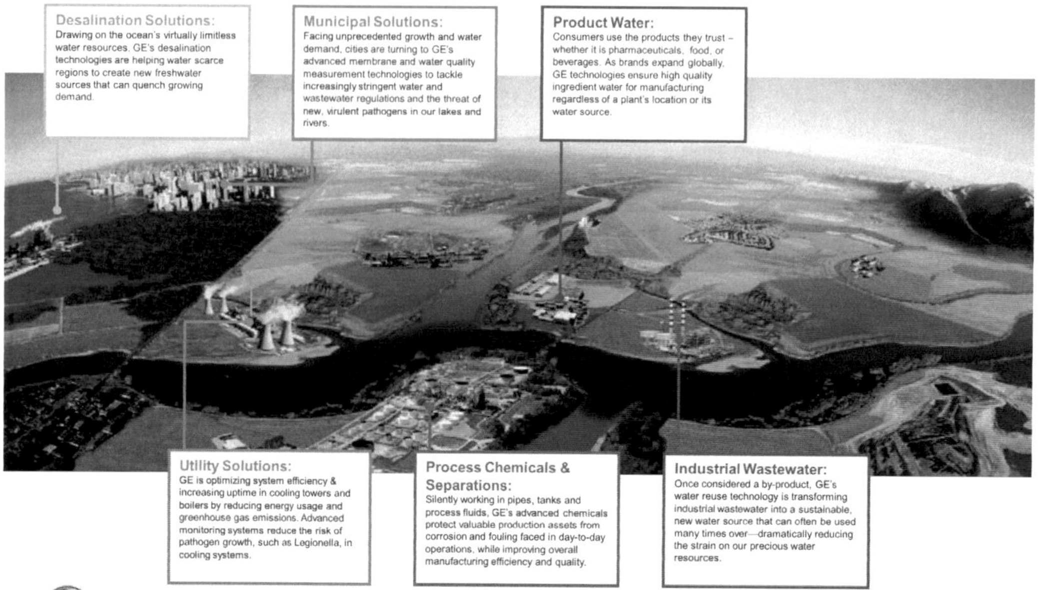

imagination at work

3

GE-Goldman Sachs-WRI Water Risk Atlas

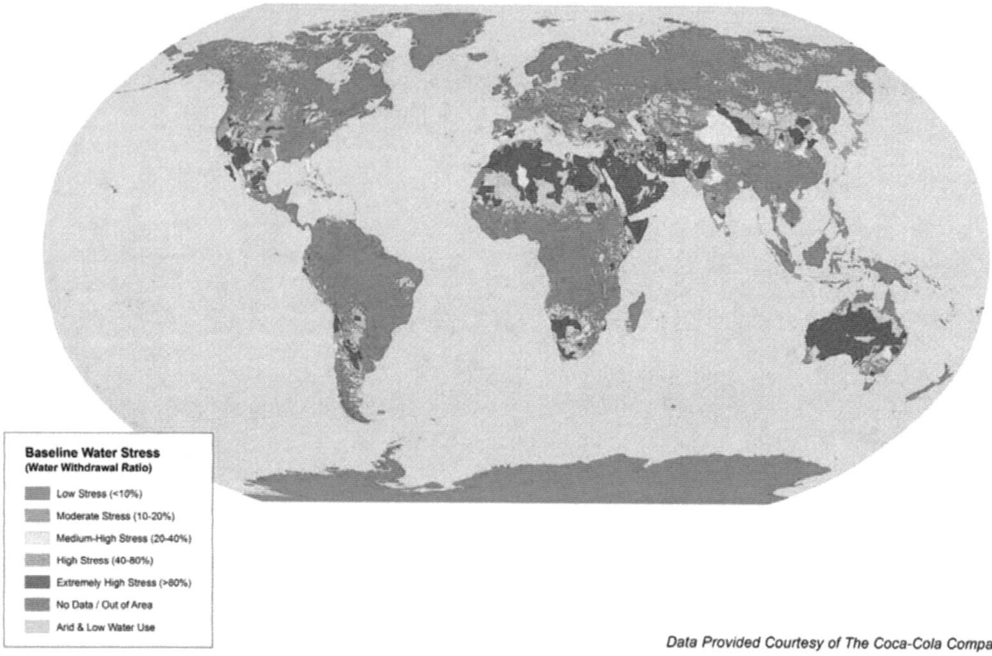

Data Provided Courtesy of The Coca-Cola Company

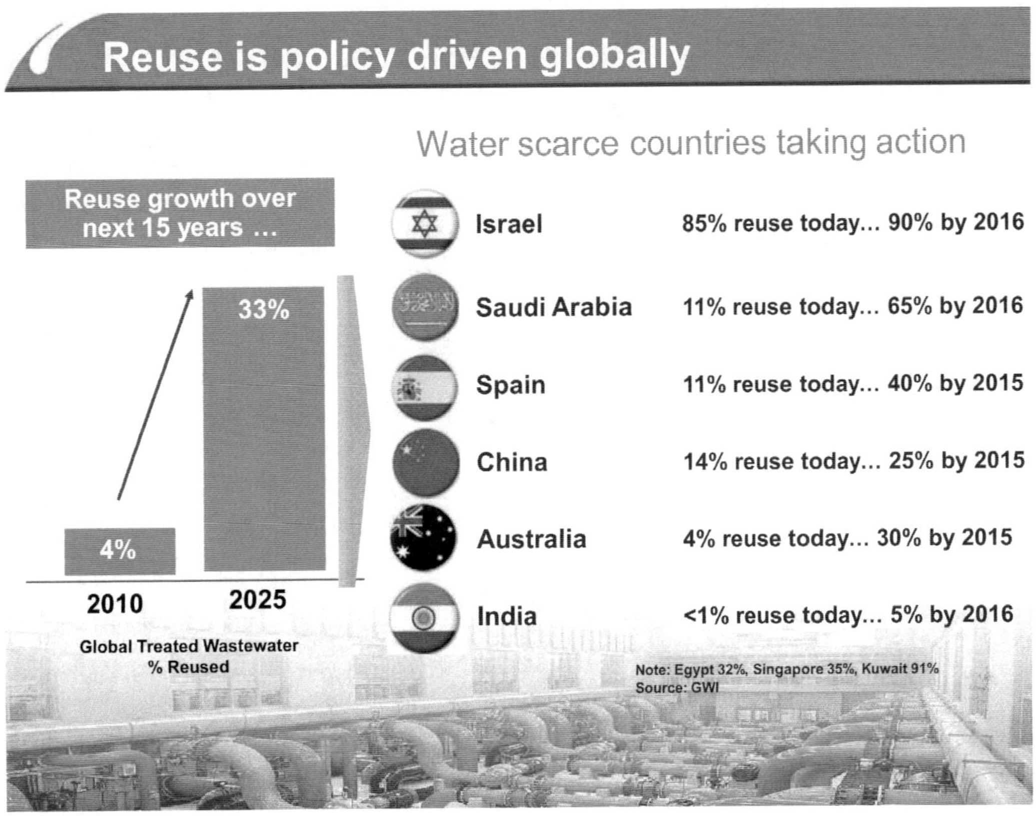
Reuse is policy driven globally
Water scarce countries taking action
Reuse growth over next 15 years ...
33%
4%
2010
2025
Global Treated Wastewater % Reused
Israel
85% reuse today... 90% by 2016
Saudi Arabia
11% reuse today... 65% by 2016
Spain
11% reuse today... 40% by 2015
China
14% reuse today... 25% by 2015
Australia
4% reuse today... 30% by 2015
India
<1% reuse today... 5% by 2016
Note: Egypt 32%, Singapore 35%, Kuwait 91%
Source: GWI

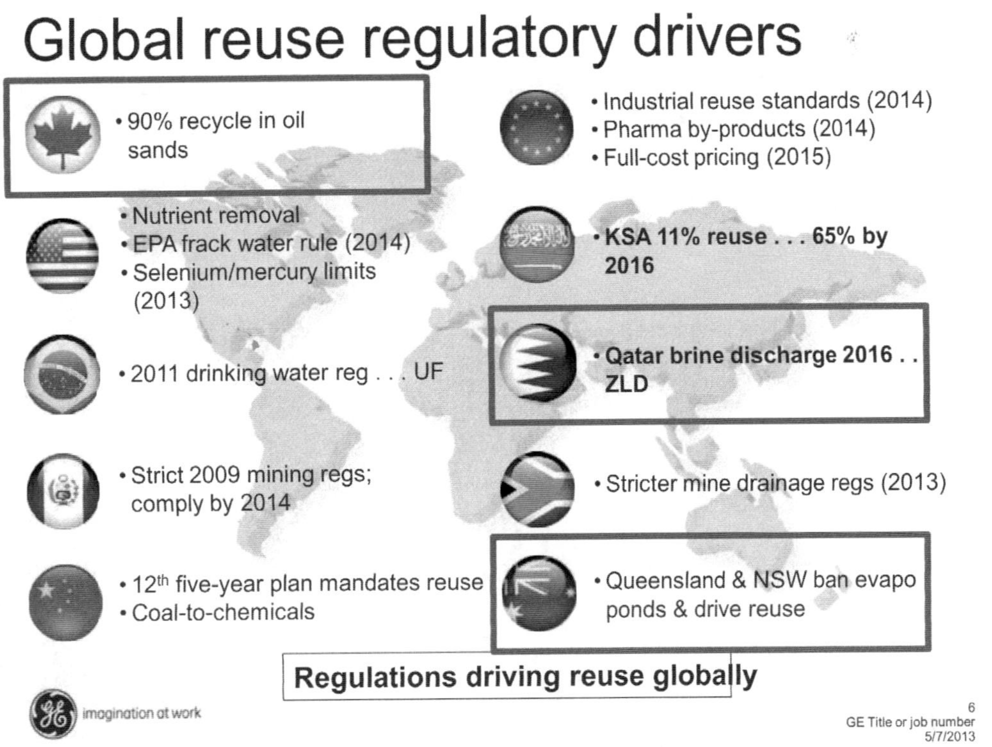
Global reuse regulatory drivers
• 90% recycle in oil sands
• Nutrient removal
• EPA frack water rule (2014)
• Selenium/mercury limits (2013)
• 2011 drinking water reg . . . UF
• Strict 2009 mining regs; comply by 2014
• 12th five-year plan mandates reuse
• Coal-to-chemicals
• Industrial reuse standards (2014)
• Pharma by-products (2014)
• Full-cost pricing (2015)
• KSA 11% reuse . . . 65% by 2016
• Qatar brine discharge 2016 . . ZLD
• Stricter mine drainage regs (2013)
• Queensland & NSW ban evapo ponds & drive reuse
Regulations driving reuse globally
imagination at work
6
GE Title or job number
5/7/2013

Water reuse legislation - 2013

- SRF Funding
- **Private Activity Bonds**
- Alternative Water Sources Act ($125/year for 3 years for Desal & Reuse)
- 10% Renewable Water Target
- Climate Change & Resilience . . . Sandy Relief
- Investment Tax Credits for Reuse
- **WIFIA**

imagination at work

imagination at work

End
imagination at work

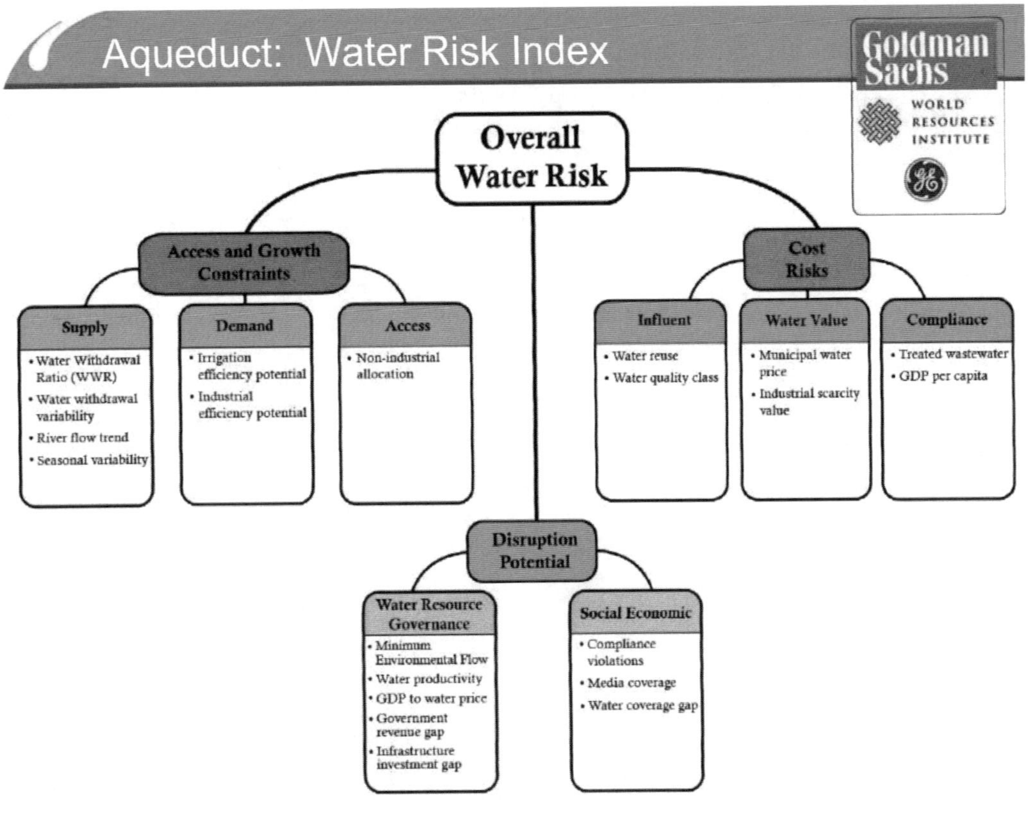
Aqueduct: Water Risk Index
Goldman Sachs
WORLD RESOURCES INSTITUTE
Overall Water Risk
Access and Growth Constraints
Supply
• Water Withdrawal Ratio (WWR)
• Water withdrawal variability
• River flow trend
• Seasonal variability
Demand
• Irrigation efficiency potential
• Industrial efficiency potential
Access
• Non-industrial allocation
Cost Risks
Influent
• Water reuse
• Water quality class
Water Value
• Municipal water price
• Industrial scarcity value
Compliance
• Treated wastewater
• GDP per capita
Disruption Potential
Water Resource Governance
• Minimum Environmental Flow
• Water productivity
• GDP to water price
• Government revenue gap
• Infrastructure investment gap
Social Economic
• Compliance violations
• Media coverage
• Water coverage gap

Global thought leader on reuse

- Global White Paper promoting greater reuse
- "A menu for policy makers"
- Downloaded >40,000 times

2011 Saudi Reuse Summit

imagination at work

11

International Society for Water Solutions
Industrial Water Use and Reuse Workshop
May 1, 2013

TEXAS REGULATORY DIRECTION FOR WATER CONSERVATION

Julie Klumpyan
Director of Government Affairs for Valero

Valero Energy

- International manufacturer and marketer of transportation fuels, other petrochemical products and power
- Valero subsidiaries employ approximately 10,500 people
- Assets include:
  - 16 petroleum refineries with a combined throughput capacity of approximately 3 million barrels per day
  - 10 ethanol plants with a combined production capacity of 1.2 billion gallons per year
  - A 50-megawatt wind farm
  - Approximately 7,300 branded wholesale outlets carry the Valero, Diamond Shamrock, Shamrock and Beacon brands in the United States and the Caribbean; Ultramar in Canada; and Texaco in the United Kingdom and Ireland.

VALERO

Agenda

- Focus on Texas Water –
  - Current status - Drought
  - Population growth impact
  - Diverse supplies - facilities must have water to operate
  - Exposure
- History of Water Planning in Texas
- Funding Options identified in Texas
- Impact of drought on current session

2

VALERO

Water in Texas – Current Status

- In 2011 – 99% of Texas is in severe, extreme or exceptional drought (Texas 2012 State Water Plan)
- Even after the rains in 2012, 82% of Texas is still experiencing extreme/exceptional drought (US Drought Monitor)
- 1000 water systems impose voluntary or mandatory water use restrictions (Texas Drought Preparedness Council)
- Some communities ran out of water (Natural Resource committee Hearings in Spring 2012 in House and Senate)
- TCEQ suspended 1,200 water rights in Brazos, Colorado, Neches and Sabine river basins for the following types of users: municipal, industrial, irrigation, recreation, domestic and livestock. (TCEQ)
- 2011 ranked as worst 1 year drought in Texas history (Texas 2012 State Water Plan)

3

Population Growth

Per Texas Water Development Board Projections

- Texas's population is projected to grow from 25 million now to 46 million in fifty years.
- The water demand is estimated to climb from 18 million acre-feet a year to 22million acre-feet a year by 2060, an increase of twenty percent.

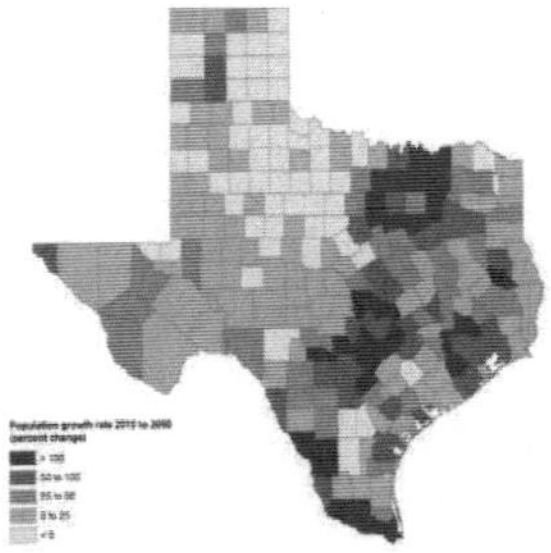

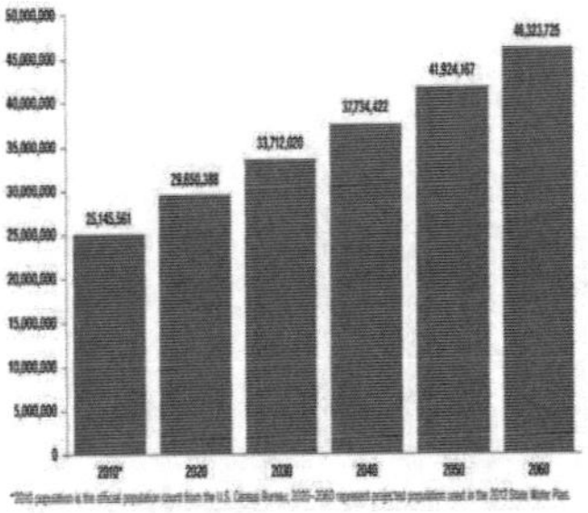

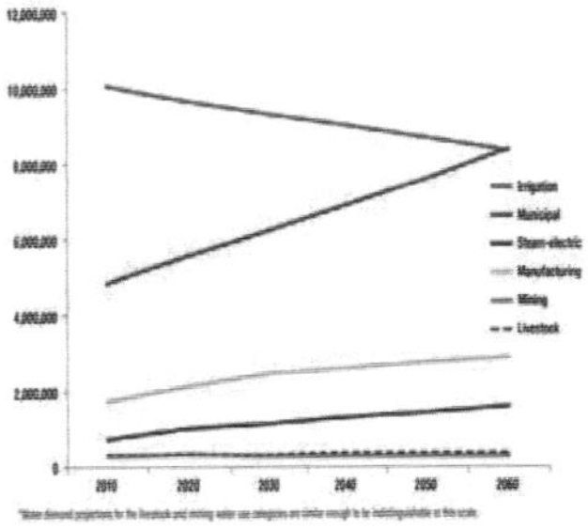

4

Diversity of Water Resources but consistent concern with lack of available water

- Water resources for our significant Texas assets:
  - Texas City purchases Brazos River water from the Gulf Coast Water Authority; this supply is threatened by residential growth
  - Corpus Christi purchases water from the City of Corpus Christi; they report they have 2 years worth of water if there is no rain, or nothing is done; expect ongoing drought is likely to require mandatory water conservation
  - Three Rivers uses city water and well water – both of which are threatened by growth
  - Panhandle uses well water but is threatened by competition from agriculture
  - Houston has indicated no immediate concerns regarding security of process water supply but need long term projects
  - Port Arthur has the largest water supply/demand, historically abundant from the canal system, but threatened restrictions
  - Headquarters is in San Antonio, which has implemented water conservation restrictions for years.
  - Dallas needs additional water due to growth

5

Exposure

- Water curtailment at facilities if water solutions are not resolved
  - **Domestic and municipal use of surface water has priority over all other beneficial uses of water** including the uses of water for industrials (Section 11.204, Texas Water Code)

- Cost of being part of the solution
  - Water efficiency projects at our plants
  - Ensure funding of Texas Water Plan

6

Texas Water Planning

- Formal water supply planning began in the 50's after the worst "drought of record" for Texas – a 7 year drought; the legislature passed the Water Planning Act of 1957, creating a water planning group
- Plans were adopted in 1968, 1984, 1990, 1992 and 1997 recognizing growth in population and the need to develop future water supplies
- In the 80's Harris County area had to get off ground water dependence and move to surface water because city was sinking; thus subsidence districts

- Due to 10 month drought in mid-90's, Lt. Gov. Bob Bullock declared that water planning would be the primary issue that session

7

More Serious Water Planning

- 1997 – Senator Brown passed SB1 requiring **consensus** based water planning by region on how to meet water needs in times of drought. Required State plan every 5 years based on the 16 regional plans. It made inter-basin transfer rights Jr. water rights. Rest of State is 1st in time, 1st in Right.
- Sen. Armbrister passed SB2 and Edwards Aquifer legislation
- Sen. Averitt passed SB3 creating process to address environmental flows, designate unique reservoir sites and sites of unique ecological value. TCEQ Sunset bill gave them rights to curtail surface water.
- Need to get started funding water plan: it's gone from a cost of $17.9B in 2002, to $31B in 2007 to $53B in 2012.

Difficulties in Funding the State Water Plan

Several sessions of FAILED proposed funding mechanisms for long term for 4 main reasons:

1. Opposition by group most heavily impacted.
   - Water rights/Conservation development fee would have exempted Ag and residential, but would have cost industry hundreds of millions a year.
   - Removal of sales tax exemption on bottled water, fought by those businesses.
2. Opposition from the entity having to collect the fee
   - Example was TX Municipal League against the tap fee
3. Exclusion of Ag from contributing, even though they use over 50% of the water in Texas
4. Legislators afraid to raise fees/taxes

Plan needs long term funding solution

- State Water Plan estimates $53 Billion needed for capital projects by 2060.
  - Going into 2013 Session, looked like we needed funding source of $150 Million a year (can be leveraged with bonds, federal funding, local funding).
    - Most traction Proposal was on electrical meters
    - While not ideal philosophically since fee on electricity and not water, lacking better annual funding solution
    - Electrics/Municipals have worked out collection allowanc

  - Texas Blessed with oil and gas boom, adding significant resources (Billions) to the Rainy Day Fund
    - Majority of legislators seem to agree that using the "rainy day fund" is the best way to finance the Texas Water Plan

10

Texas Legislative Solution-2 major parts

- Part 1 is HB4 by Chairman Ritter and SB4 by Chairman Fraser
- Creates State Water Implementation Fund for Texas (SWIFT) – water infrastructure bank that operates as a revolving account for financing projects included in the State Water Plan
  - Support thru state funding of locally identified water projects
  - Funds lent through SWIFT would be repaid and made available for future local and regional water projects
  - $2 Billion investment in SWIFT would satisfy the $27 billion in state-assisted financing requested by local planning groups implementing projects over next 50 years.

11

Funding Mechanism – Part 2

- Several bills filed in House and Senate
- Allocate $2 Billion from Rainy Day Fund as one time funding source
- No significant opposition to water funding
- Disagreement is in whether to take it to the voters to avoid the Constitutional spending cap
- Prognosis for passage of water funding is better than it has ever been

12

Time to $Fund the Texas Water Plan

- Texas Legislature rises to the occasion when issues must be resolved: Worst 1 year drought in history in 2011 covering 99% of the State, costing Billions of dollars
- Funding the State Water Plan is not a silver bullet solution
- Several proven water tools will be necessary
  - Conservation
  - Surface Water projects
  - Groundwater projects
  - Reuse and other

13

Water Management Tools that work

See H2O4TEXAS.ORG

- Water conservation will still be necessary thru-out Texas
  - Efficiency of use & decreased demand on existing supplies
  - Savings from every sector including Industrial Users: Ensure we are not viewed as water hogs; we must demonstrate water efficiency
- Surface Water
  - Connection to existing supplies via pipelines and other tools
  - Building/expanding water treatment plants
  - Stream diversions
  - Reservoirs
- Groundwater
  - New Wells
  - Aquifer Management

14

Water Management Tools that work

See H2O4TEXAS.ORG

- Reuse – treated wastewater for irrigation, etc.
- Treatment of Brackish Water
- Desalinization
- Conjunctive use of surface and groundwater

15

VALERO

Do your part!

THANK YOU

16

8
Hu
HUMAN

Today's Agenda

- Dow's commitment and societal challenges
  - 2015 goals
- Manufacturing Operations: Water Footprint Management
  - Case Histories
- Dow and The Nature Conservancy Collaboration
  - Preliminary Results

8
Hu
HUMAN

Our Water Vision

Dow will be a leader in addressing the global water crisis by setting the standard for sustainable water use and management. We will develop innovative technologies and business models that lower the cost of water purification, set new levels for efficient water use at our manufacturing facilities while striving to reach zero water discharge. Dow will use creative partnerships to increase the global sense of urgency to solve this issue.

2015 Sustainability Goals*

- Our goals were inspired by the UN Millennium Development Goals and are consistent with the science and technology we do best. They will drive our growth profitability, continued EH&S performance improvements, and a focus to help address some of the world's most pressing challenges.

- The seven 2015 Goal components
  - **Local Protection of Human Health & Environment**
  - Contributing to Community Success
  - Product Safety Leadership
  - Sustainable Chemistry
  - Breakthroughs to World Challenges
  - Energy Efficiency and Conservation
  - Addressing Climate Change

* www.dow.com

Science and Technology Company

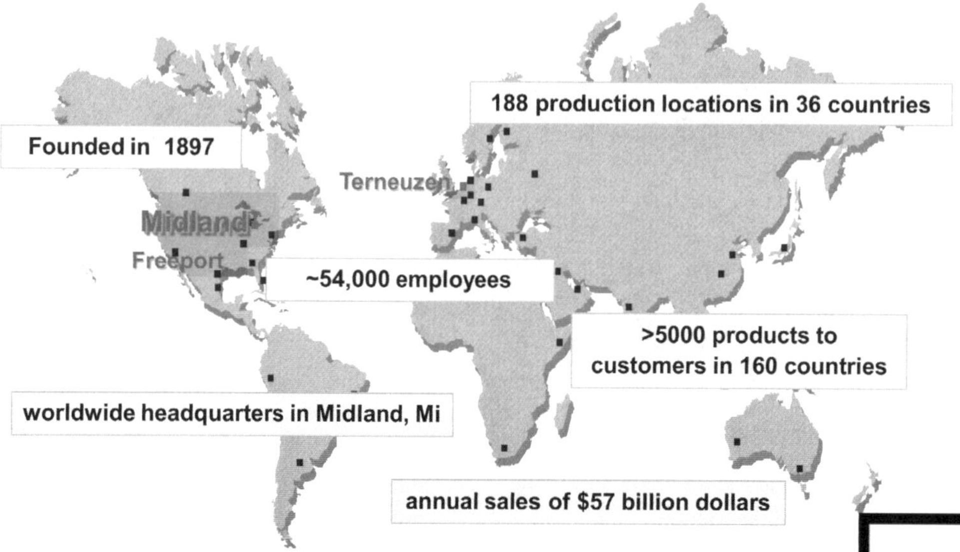

You find our chemicals everywhere

- Food
- Building Maintenance & Construction
- Transportation
- Furniture & furnishings
- Paper and Publishing
- Home Care Improvement
- Sports
- Personal and Household Care
- Health and Medicine
- Water Purification
- Electronics and Entertainment

Water is …

Regional

Local

Abundant… but not available

Different drivers require
Different solutions

One size does not fit all…

8
Hu
HUMAN

Water Availability

- Address on multiple fronts
- Technology
  - solution provider
  - solution implementer
- Recycle/reuse
- Operations Excellence
- Partnerships / Relationships

8
Hu
HUMAN

Dow Water & Process Solutions

- Business unit comprised of component technologies designed to advance the science of water purification around the world
- Target applications are seawater desalination, contaminant removal and water reuse.
- FILMTEC™ reverse osmosis membranes, DOWEX™ ion exchange resins, ADSORBSIA™ GTO™ titanium-based arsenic removal media, ultrafiltration, and electrodeionization

Technological Process Expertise and Innovative Solutions

- At BASF's site in Belgium, Dow and BASF jointly developed an innovative hydrogen peroxide to propylene oxide (HPPO) technology, which provides economies of scale, and reduces wastewater and energy usage.
  - New plants built using this technology:
    - receive a reduction in wastewater by more than 70 percent
    - energy usage reduced by 35 percent when compared with PO process technologies in use today.

- Dow's largest manufacturing site globally in Freeport, Texas, has designed and implemented a power plant using seawater for cooling instead of freshwater.
  - Seawater use saves $600,000 / year in avoided treatment, energy and water costs.
  - Resulted in 1,200 gallons per minute reduction in freshwater demand and this means that there will be 1,200 gallons per minute reduction for the life of this facility.

8
Hu
HUMAN

Recycle / Reuse

- Stade Germany Site
  - Vigorous R&D and project implementation
  - Recycle of the wastewater generated from the propylene oxide production, after biological and physical/chemical treatment, back to the brine mining operation
  - Results is 10 million cubic meters per year of freshwater conserved and correspondingly, about 10 million cubic meters per year of wastewater not discharged.
- ***More on Recycle/Reuse***

8 Hu HUMAN

Water Footprint Management Manufacturing Operations Case Histories

8 Hu HUMAN

Water management

The 3X Water Model

Dow Terneuzen

Reusing Household Waste Water Through a Public Private Partnership

8 Hu HUMAN

The 3X Water Model

- **Closing the water cycle developed and implemented in the region Terneuzen:**
- Reduce pollution at source (in-process)
- Classic reuse (return steam condensate)
- Reuse industrial wastewater
- Regional integration by reuse of municipal wastewater for industrial application
- Utilize alternative sources in the region

8 Hu HUMAN

Background

- The region suffers from a structural lack of sweet fresh water
- The Dow site in Terneuzen, the Netherlands, is the largest single water user in the region; it uses approx. 60,000 m^3 water per day
- Dow strives to constantly improve water efficiency
  - reduce fresh water use
  - maximise re-use
- Dow's focus is to not use scarce fresh water sources which are suitable for drinking water
- Dow and Water supplier Evides pro-actively engaged with the city of Terneuzen and the Water Board to manage the regional water balance

8 Hu HUMAN

The 3X Water Model - A successful PPP -

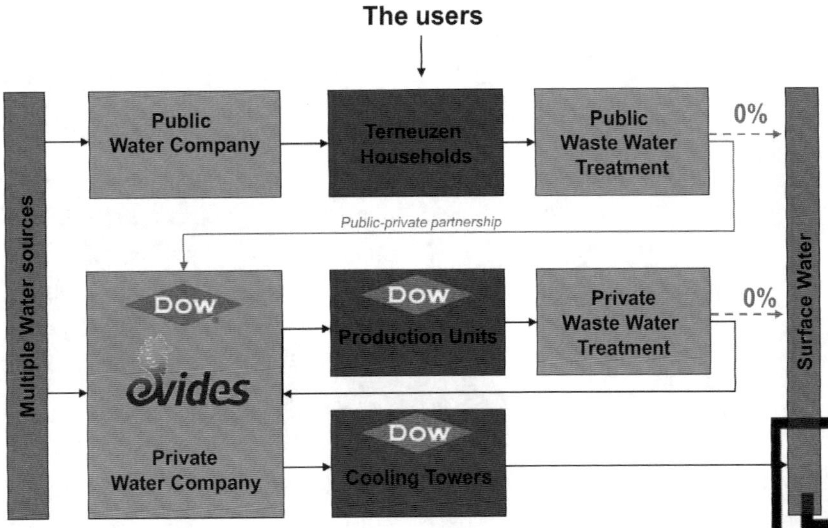

8 Hu HUMAN

Data Dow Terneuzen

- **Daily water usage**
  - Dow location ~ 60,000 m^3 (excluding seawater and potable water)
  - Fresh water demand 22 million m^3/yr
  - 5 x regional consumption in households
- **Realized Reuse**
  - Classic reuse of condensed steam 6 million m^3/yr
  - Dow wastewater 3 million m^3/yr
  - City of Terneuzen wastewater 1.5 million m^3 /j
  - Rainwater 0.5 million m^3/yr
- **Added Value**
  - Energy saving vs. seawater evaporative desalination (60 kton/yr CO2)
  - Less chemicals 500 ton/yr
  - Reduced discharge to receiving water body
- **Present fresh water intake 11 million m^3/yr**
- **Goal for 2015 is 7 million m^3/yr (or less ...)**

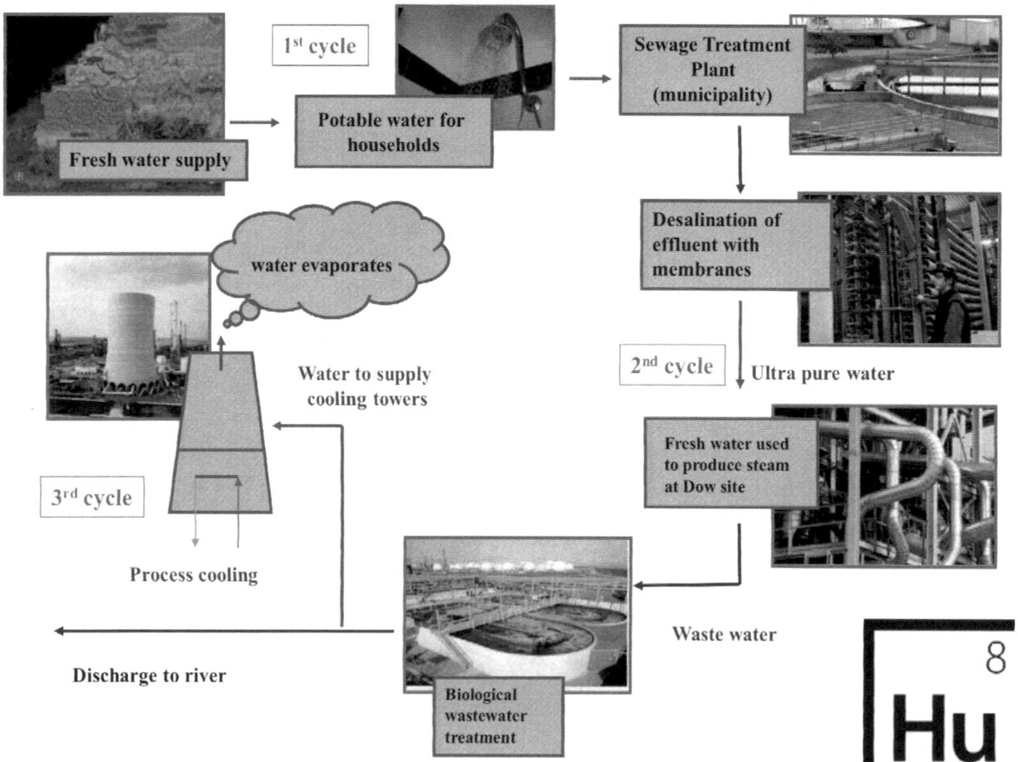

Next Steps

Terneuzen

- PPP Phase 2: building of a new membrane installation at public treatment facility:
  - Increased volume of water available for the 3X model
  - Improvement of the pretreatment in the Dow/Evides facility
  - Increased capacity of the public water treatment facility

- A study has been initiated into additional sources of fresh water which after treatment can either feed into the 3X model or into agricultural sector

Global

- Leverage (parts of) the 3x model to other Dow locations
- Incorporate the concepts in new facility design
- Development of components that will allow the reuse of cooling water

8
Hu
HUMAN

Largest chemical manufacturing facility in the world

8
Hu
HUMAN

Dow Freeport Burning Platform

SEAWATER INTRUSION

Reality

•Previous Strategy → Interruptible Water

PERCEPTION: THE BRAZOS RIVER HAS LARGE FLOWS RELATIVE TO DOW/LOCAL DEMAND

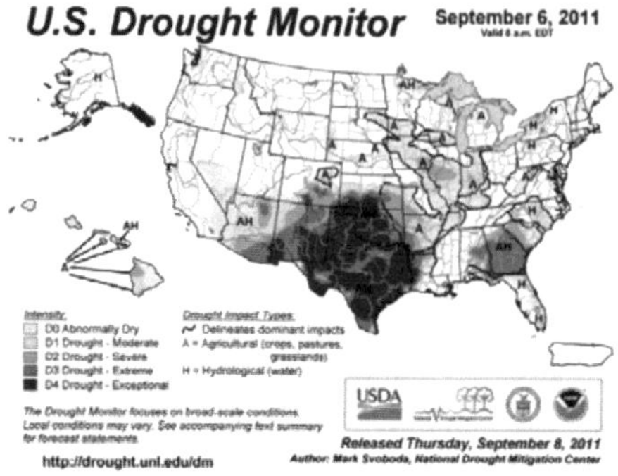

8 Hu HUMAN

•Dow Freeport Water Infrastructure

•Dow Intakes, Reservoirs and Canals

•8-12 days travel time from non-Dow Reservoirs

•Harris Reservoir (River Mile 46)
•7000 Acre-Ft of Storage
•Available Storage = 14 Days

•Brazoria Reservoir (River Mile 25)
•21,700 Acre-Ft of Storage
•Available Storage = 30 Days

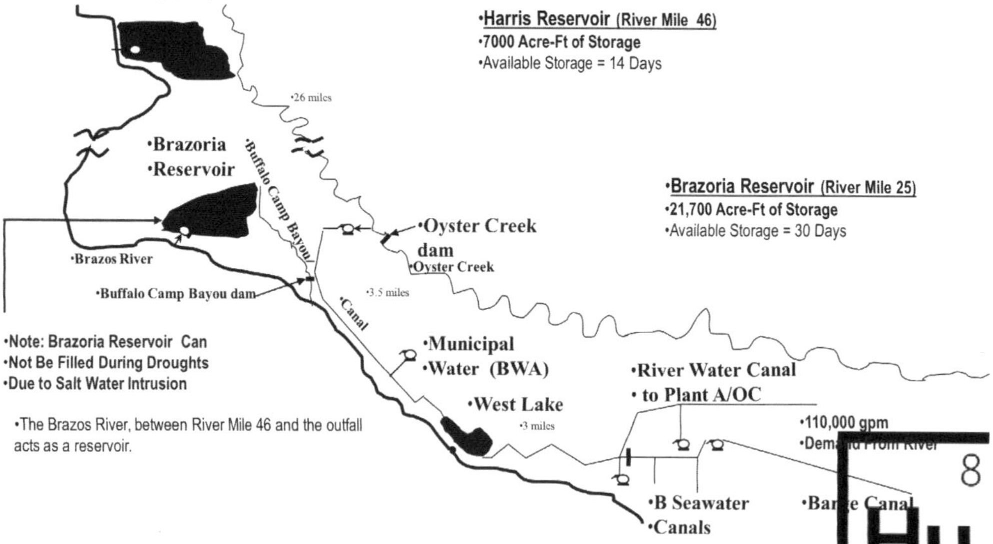

•Note: Brazoria Reservoir Can
•Not Be Filled During Droughts
•Due to Salt Water Intrusion

•The Brazos River, between River Mile 46 and the outfall acts as a reservoir.

8 Hu HUMAN

Call to Action

- Water Summit in November 2011
- Initiated site contest to identify water conservation opportunities – raise awareness and tap into the brainpower of all employees – September 2011
- Hosted a Water symposium in May 2012 with the Brazoria County Petrochemical Council

8
Hu
HUMAN

Water Summit November 2011

- 60+ Dow water and non-water experts
- **Outcome**
  - 276 ideas to address:
    - Demand Reduction
    - Improved Storage
    - Long Term Secure Supply
  - **Water Strategy Team Formed**

8
Hu
HUMAN

Texas Operations Water Strategy

Element 1 - Conservation

Water is not free or infinitely available, and must be managed as a valuable resource in the future.

Element 2 – Security of Supply

A "Secured" Water Supply that Dow can control or influence is critical for business operations.

8
Hu
HUMAN

What is Dow doing?

STORAGE AND PUMPING CAPACITY

- Acquire land for a reservoir expansion project (2011)
- Add pumping capacity
- Design and permit reservoir

REDUCE INTERNAL DEMAND

- 10% reduction achieved 4th qtr 2011
- An additional 10% reduction achieved in 2012/early 2013
- Change the way we think about water

DETERMINE STRATEGIC DIRECTION

- New Facilities
- Alternative Water Technologies

8
Hu
HUMAN

Harris Reservoir Expansion Plan

- Sought legislative action 2011 to make land available for sale

> - **Dow gets prison land**
> - THE FACTS - Dow Chemical Co. will deepen the pool of water resources for the company's Freeport site and southern Brazoria County residents after a $5.5 million bid won the company a 2,200-acre parcel of former prison land on Thursday…

Figure 1-3. Harris Reservoir with Pre-sedimentation Basin and Brazos River Intake and Pump Station in Foreground

Figure 2-2. North Levee of Harris Reservoir with Prison Farm Land to the Right

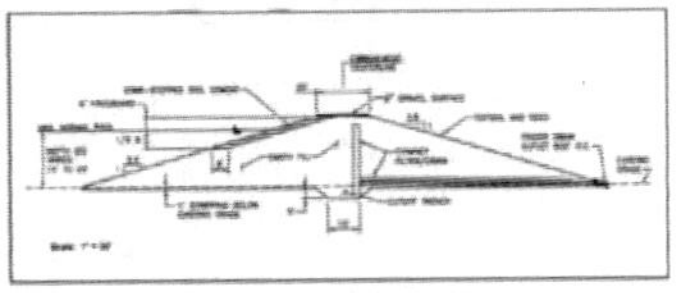

Figure 4-1. Harris Reservoir Expansion Project, New Reservoir Alternative, Conceptual Embankment Section

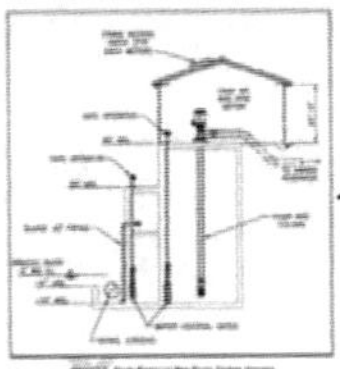

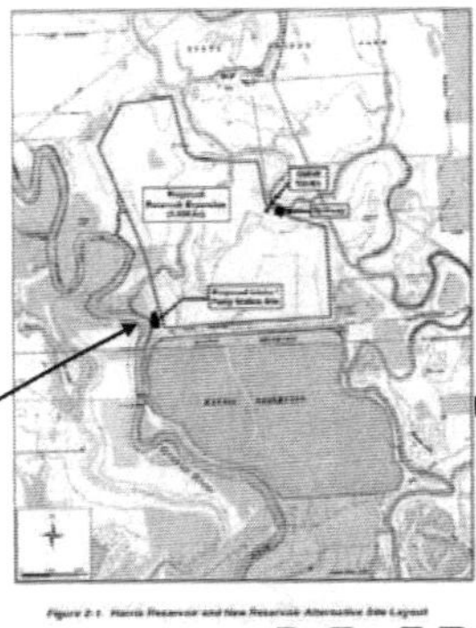

Figure 2-1. Harris Reservoir and New Reservoir Alternative Site Layout

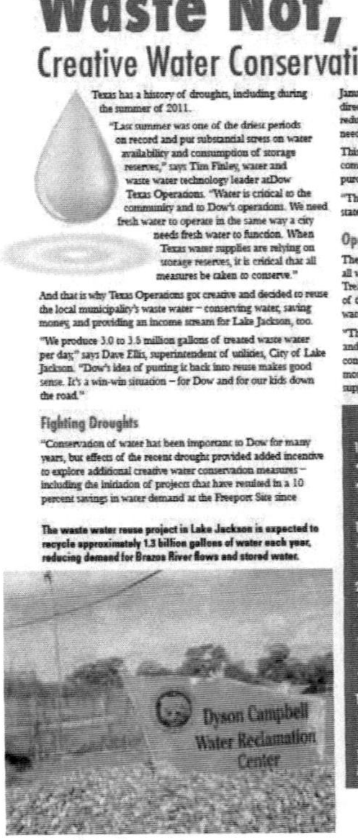

Waste Not, Want Not

Creative Water Conservation Makes an Impact

Texas has a history of droughts, including during the summer of 2011.

"Last summer was one of the driest periods on record and put substantial stress on water availability and consumption of storage reserves," says Tim Finley, water and waste water technology leader atDow Texas Operations. "Water is critical to the community and to Dow's operations. We need fresh water to operate in the same way a city needs fresh water to function. When Texas water supplies are relying on storage reserves, it is critical that all measures be taken to conserve."

And that is why Texas Operations got creative and decided to reuse the local municipality's waste water – conserving water, saving money, and providing an income stream for Lake Jackson, too.

"We produce 3.0 to 3.5 million gallons of treated waste water per day," says Dave Ellis, superintendent of utilities, City of Lake Jackson. "Dow's idea of putting it back into reuse makes good sense. It's a win-win situation – for Dow and for our kids down the road."

Fighting Droughts

"Conservation of water has been important to Dow for many years, but effects of the recent drought provided added incentive to explore additional creative water conservation measures – including the initiation of projects that have resulted in a 10 percent savings in water demand at the Freeport Site since January 2011," says Doug Whipple, water strategy project director. "Every gallon of fresh water demand that can be reduced lessens the draw on local reservoir levels and reduces the need to pump water from the Brazos River.

This is particularly important when Dow and the local communities are in a low-water situation and need to request or purchase water from upstream reservoirs.

"This is a beneficial reuse of water. It benefits the city, Dow, the state…everyone!" says Dave.

The waste water reuse project in Lake Jackson is expected to recycle approximately 1.3 billion gallons of water each year, reducing demand for Brazos River flows and stored water.

Opening the Valve

The 2011 drought helped launch the waste water reuse project – all with the opening of a valve at the Lake Jackson Waste Water Treatment Plant on November 1, 2011. Since then, 100 percent of the city's treated waste water has been diverted to Dow's fresh water canal for use in its manufacturing processes.

"The project demonstrates the collaboration and effort of Dow and the local community to go to extensive measures to be water conscious," says Tim. But this isn't the end. "There are many more collaborative projects needed statewide to protect water supplies and stored water reserves in times of drought."

A History of Drought

1950s - Drought
- Dow installs local reservoirs
- Water-supply contract established for existing/future reservoirs in Brazos Basin

1990s – Population Growth
- Growth – both locally and in the entire Brazos Basin – adds stress on water reserves in reservoir storage

2000s – Water Conservation
- Dow begins projects to conserve water/ improve water supplies capabilities
- Extensive studies completed to better understand reliability of local reservoirs and Brazos Basin reservoirs to meet demand
- As a result of this work, Dow procures land for additional local storage

Today – Waste Water Recycling
- Freeport site adds waste water reuse program with the City of Lake Jackson

You can make a difference, too! For tips on conserving water, visit http://www.wateruseitwisely.com/100-ways-to-conserve/index.php

SPRING/SUMMER 2012 • 19

- Waste water reuse project in Lake Jackson
- Expected to recycle approximately 1.3 billion gallons of water each year
- Reducing demand for Brazos River flows and stored water.

2011 Conservation Action

Conservations Projects

- Lake Jackson Water Recycle 2500 gpm
- Chlorine 4 recycle 2000 gpm
- Cooling Tower Optimization 1000 gpm
- "As Needed" conservations measures 3500 gpm

Total 2011 Impact 10,000 gpm

8
Hu
HUMAN

Freeport 2012 Site Water Reduction Efforts

Project	Status	Maximum gpm	% Site Demand	% River Demand
Enable a process technology Soft Water Recycle (Contingent plan)	July, 2012	5,000	6.60%	5.00%
Demin. Water Resin	August, 2012	1,600	2.10%	1.60%
Recycle Projects	December, 2012	5210	7.00%	2.70%
Salt Recycle Opportunity	Completing Feasibility	1,000	1.30%	1.00%
Additional cooling water system optimization	Scoping	2,500	2.60%	2.00%

Total Water Demand Reduction (Max.) 10,710 14.1% 10.7%

Water reduction with contingent option 15,710 20.7% 15.7%

- Freeport Site Water Demand
- Total from River
- = 100M gpm
- Net to Freeport Site After
- BWA and Fence Line Industry
- = 76M gpm

8
Hu
HUMAN

The Dow Chemical Company Texas Operations Received

2013 Texas Environmental Excellence Award – Water Conservation

8
Hu
HUMAN

The Economics of Ecosystems

The collaboration between Dow and The Nature Conservancy demonstrates that protecting nature can be

- both a global business strategy and
- a company priority

By combining our resources and expertise, we are integrating the value of nature into Dow's business decision-making

8
Hu
HUMAN

Freeport Freshwater Pilot Selected

The Brazos River - primary source of freshwater to Dow's Freeport Site

- critical for the site operations as well as for other water users and wildlife in the water basin.

The State of Texas Region H Water Plan forecast the water demand, on the Brazos, to increase 54% by 2060.

- Municipal and industrial demands are the primary contributors to the demand increase.

Trend associated with climate models predict increased average temperatures along with increased water loss from evaporation

- which if occur, will affect volume of flow in the Brazos

Early Work

Collaboration scientists incorporated the effects of predicted climate change and increase demand into existing models of the Brazos River basin to predict future water availability.

- Previous analysis had not considered these synergistic effects.

The models predict that there will be:

- more frequent and severe basin-wide water shortages
- with the maximum length of shortages in the lower basin where Dow withdraws its water potentially increasing three-fold

Analysis confirmed scenarios of potential reduction of future freshwater flows in the Brazos :

- studies conducted independently by Dow using historic flows,
- while adding new information about the influence of predicted higher temperatures

Stepping Into the Cost Side

Traditionally,

- companies may have projected th... f water.
- Disruption to...

Ecosystem service models can help better predict the frequency of these drought events, improving financial models.

...pay much more than ...pply is not disrupted.

Mu...

...en available (costs of water rights) but ...ost during water interruptions (up to and including cost of desalination),

- associated loss impacts must be factored into cost comparisons.

Macro Results

Nature-based and collaborative solutions may provide cost-competitive ways to improve the value of Dow's freshwater assets incrementally, while also benefiting other users and ecosystems.

Collaboration experts brainstormed potential response initiatives, which were narrowed to the following five that were analyzed in more detail:

- Wastewater reuse – construct wetlands to filter purchased municipal wastewater
- Reservoir flood pool reallocation – support the reallocation of flood pool storage at U.S. Army corps of Engineers' reservoirs and complementary floodplain restoration
- Panhandle land management – replace high water-use invasive plants with lower water-use native plants in the Texas panhandle
- Municipal Rebate – support re bate program for municipal users to buy water efficient appliances or switch to native landscaping that does not require watering
- Agriculture Water Reallocation – support for fund improvements in irrigation technology and purchase saved water

Results (cont.)

All response initiatives, except for wastewater reuse, were found to be cost-competitive with the base solution of expanding Dow's existing reservoir system.

- Potential to enhance basin supplies of water by 60,000 acre-ft per year.
- These types of solutions could complement traditional engineering approaches.

Initiatives also provide additional benefits to the public and ecosystems.

- Collectively these initiatives could protect 21,500 acres of habitat and generate public benefits of at least $10 million over the next 30 years.
- However, these actions have legal, political and technical challenges, which would need to be addressed before implementing.

Companies are beginning to shift their thinking about water.

- Previously considered an almost limitless, free or low-priced commodity
- Now recognized as a finite resource that needs to be considered when making strategic decision.
- The importance of understanding the "economics of water" will continue to i as higher demand and reduced supply lead to water scarcity in varying region around the world.

Working Together

- Private and public sectors must work together
- Solution goes beyond just technology
- Government must drive integrated policies
- Society must conserve and reuse supplies
- Businesses must create technology solutions that make economic sense and solve problems
- Partnering to solve the world's water crisis is possible and is happening as we speak

DOW

8
Hu
HUMAN

NewPage®

Water Use and Reuse in an Intergrated Paper Mill

Carson Barry – NewPage Corporation

2011 Board and Paper Production by Region*

NewPage®

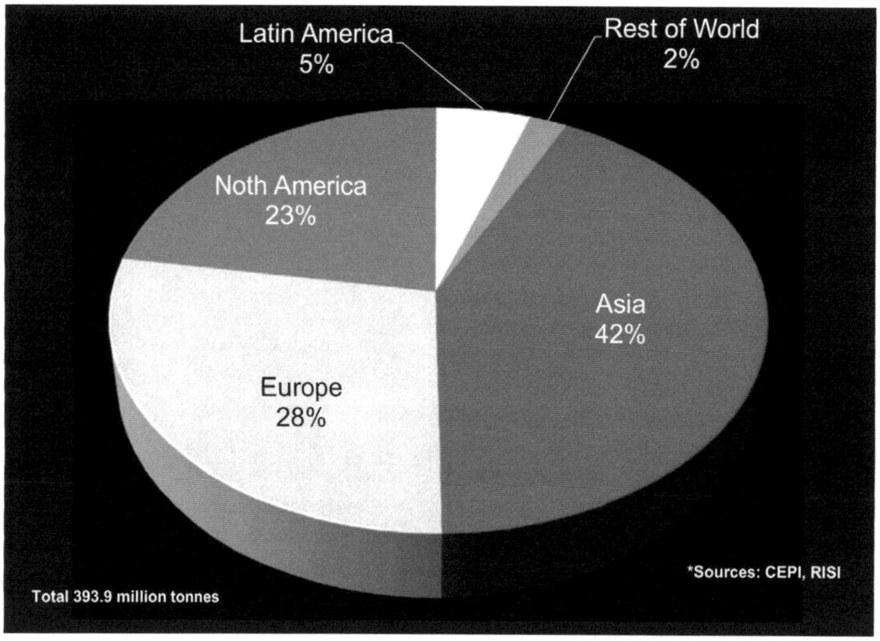

2

Global Paper Consumption Trends

http://environmentalpaper.org/documents/state-of-the-paper-industry-2011-full.pdf

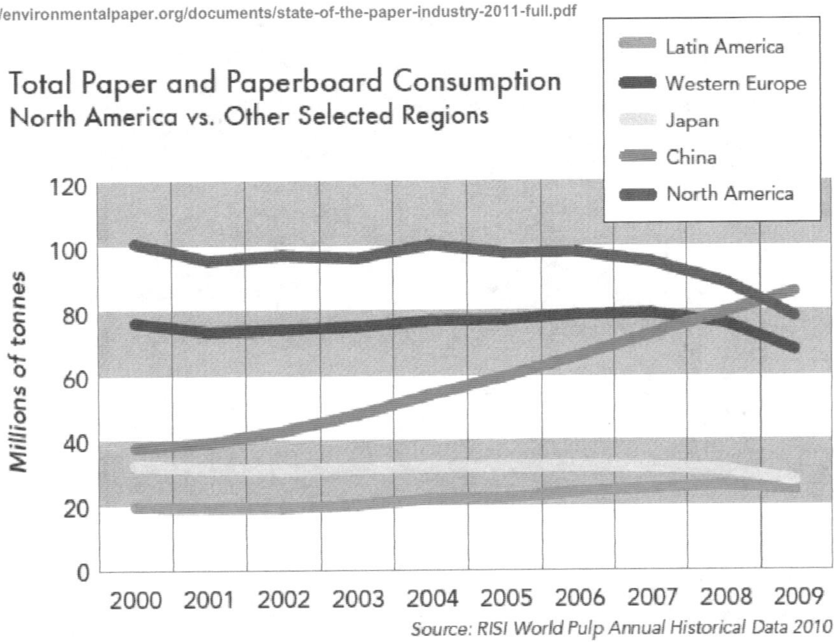

3

US Paper Segment Breakdown (Tonnage)

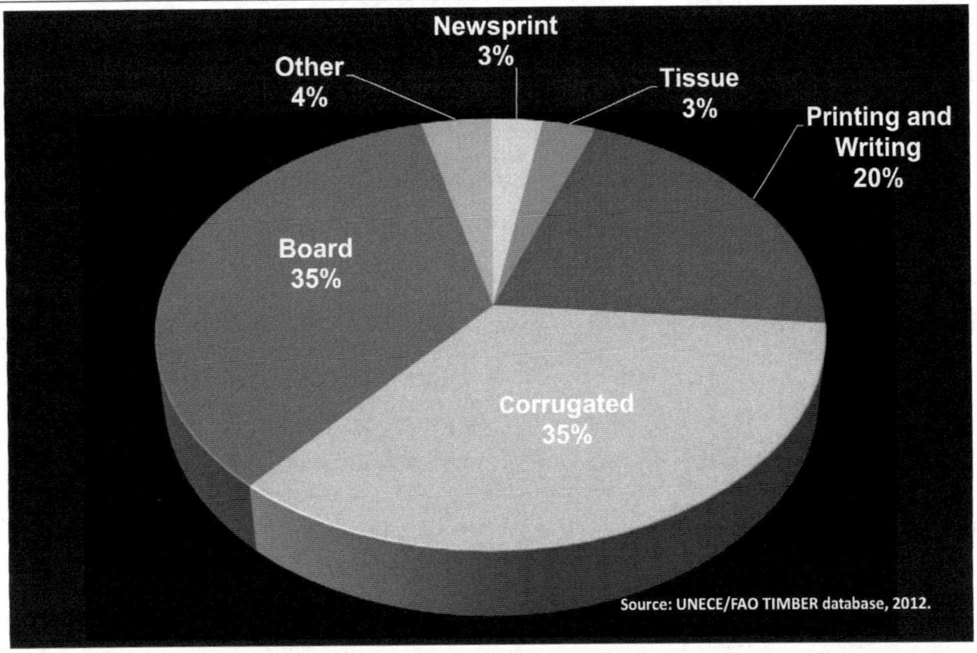

4

US Pulp and Paper Mills

NewPage

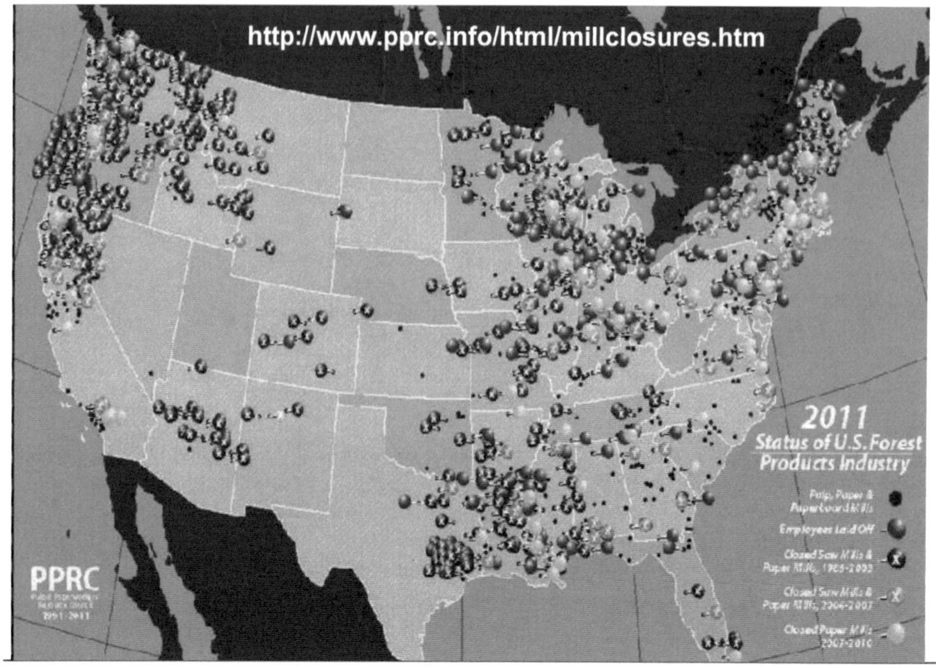

5

Canadian Pulp and Paper Mills

NewPage

6

US Paper Industry Stats

NewPage

EMPLOYMENT	
Forestry & Logging	107,100
Wood Products	388,600
Pulp & Paper	391,400
Total Employment	**887,100**
ANNUAL PAYROLL INCOME *(in thousands of dollars)*	
Forestry & Logging	$3,073,000
Wood Products	$16,501,000
Pulp & Paper	$29,911,000
Total Compensation	**$49,485,000**
NUMBER OF MANUFACTURING FACILITIES	
Sawmills, Millwork, Treating	674
Engineered Wood and Panel Products	186
Other Wood Products	125
Total Wood Products	985
Pulp, Paper & Paperboard Mills	367
Converted Paper Products	4492
Total Paper Manufacturing	4859
Total All Segments	**5,844**

Source: AF&PA as of Dec/2012

7

Paper Mill Schematic

NewPage

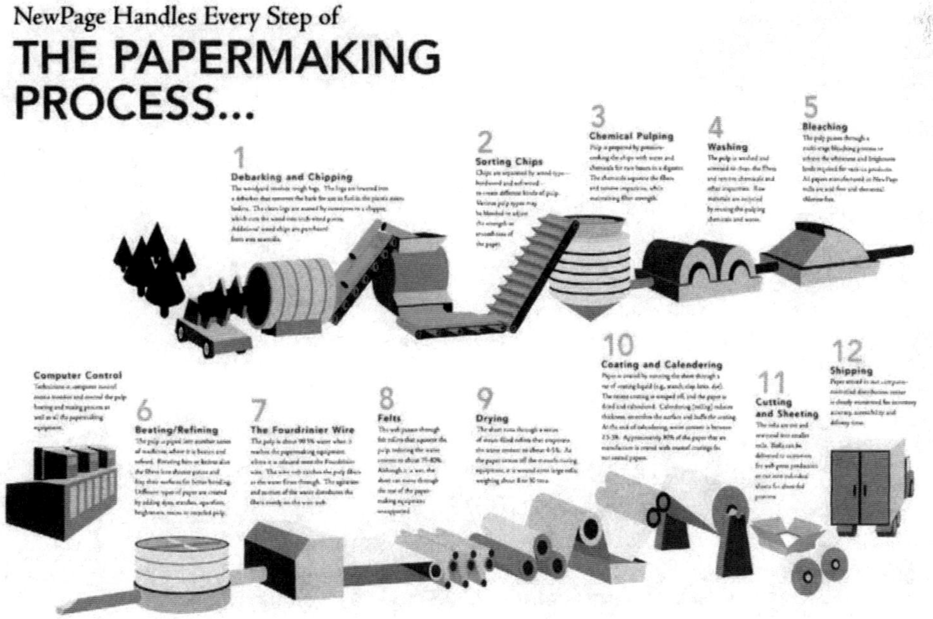

8

Water Use and Reuse In Papermaking Overview

NewPage

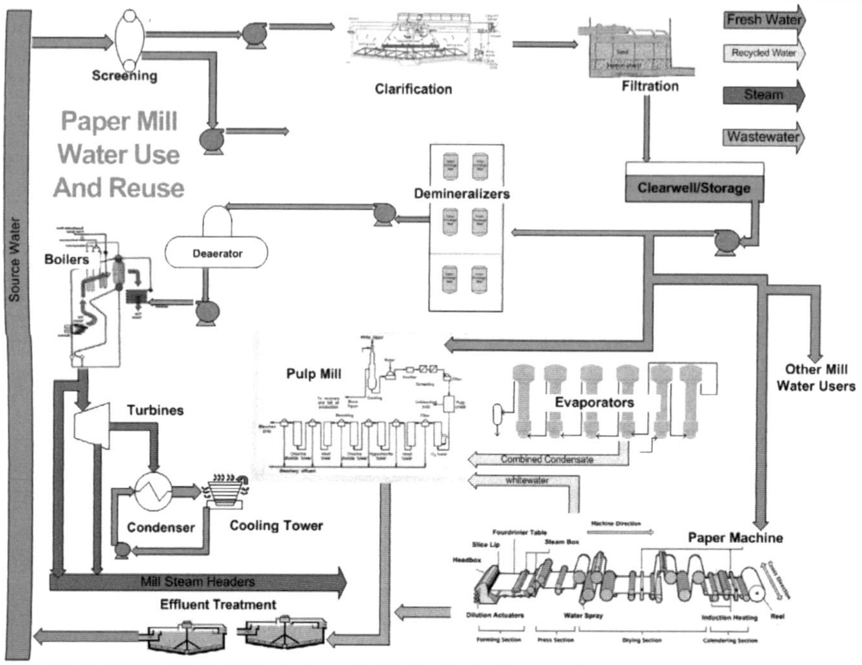

9

NewPage

Paper Mill
Major Water Recycle Streams

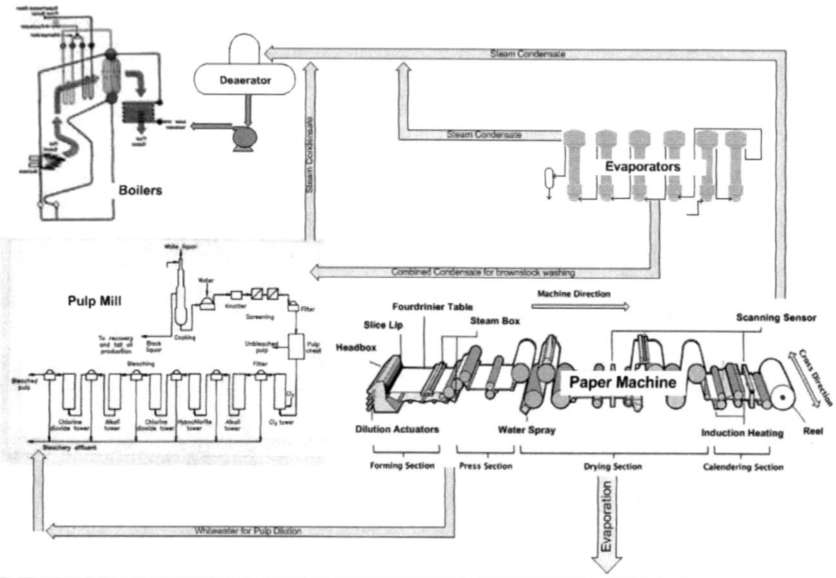

10

NewPage®

Millions of gallons per day

Fiber = $

Coating = $

Organics

Heat = $

Water = $

NewPage®

Thank You

Questions?

VALERO

Water Reuse/Recycle...What does it mean for Refiners?

Industrial Water Use and Reuse Workshop

April 30 – May 2, 2013

VALERO

Safe Harbor Statement

Statements contained in this presentation that state the Company's or management's expectations or predictions of the future are forward–looking statements intended to be covered by the safe harbor provisions of the Securities Act of 1933 and the Securities Exchange Act of 1934. The words "believe," "expect," "should," "estimates," and other similar expressions identify forward–looking statements. It is important to note that **actual results could differ materially from those projected in such forward–looking statements. For more information concerning factors that could cause actual results to differ from those expressed or forecasted, see Valero's annual reports on Form 10-K and quarterly reports on Form 10-Q**, filed with the Securities and Exchange Commission, and available on Valero's website at www.valero.com.

2

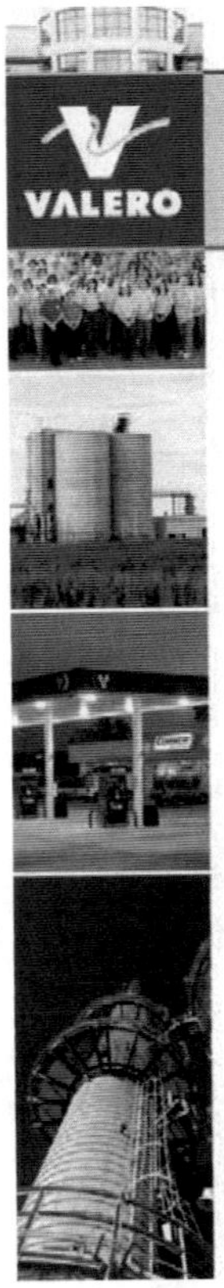

Valero Energy Overview

- **World's largest independent refiner**
  - 16 refineries
  - 3 million barrels per day (BPD) of capacity, with average capacity of 190,000 BPD (187,000 BPD without Aruba)
- **Approximately 7,300 branded marketing sites**
- **One of the largest renewable fuels companies**
  - 10 efficient corn ethanol plants with total of 1.2 billion gallons/year (72,000 BPD) of nameplate production capacity
    - All plants located in resource-advantaged U.S. corn belt
  - 50 MW wind farm in Texas
  - Diamond Green Diesel JV under construction (renewable diesel from waste cooking oil and animal fat)
    - 10,000 BPD capacity, 50% to Valero
  - Portfolio of investments in next-generation fuels
- **Approximately 10,500 employees**

Valero's Geographically Diverse Operations

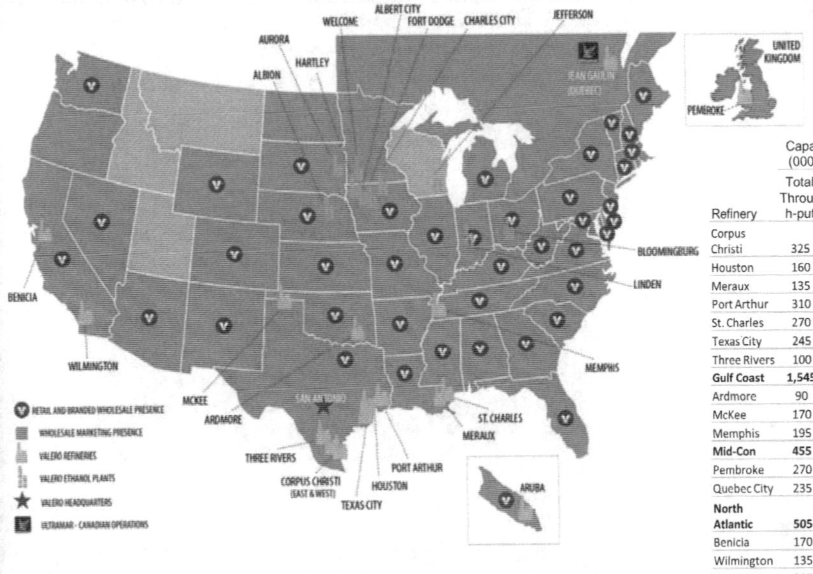

Shutdown in March 2012
235,000 bpd capacity Nelson Index of 8

Refinery	Capacities (000 bpd) Total Throug h-put	Crud e Oil	Nelso n Index
Corpus Christi	325	205	20.6
Houston	160	90	15.1
Meraux	135	135	10.2
Port Arthur	310	290	12.7
St. Charles	270	190	15.2
Texas City	245	225	11.1
Three Rivers	100	95	12.4
Gulf Coast	**1,545**	**1,230**	**14.0**
Ardmore	90	86	12.0
McKee	170	168	9.5
Memphis	195	180	7.5
Mid-Con	**455**	**434**	**9.2**
Pembroke	270	220	11.8
Quebec City	235	230	7.7
North Atlantic	**505**	**450**	**9.7**
Benicia	170	145	15.0
Wilmington	135	85	15.8
West Coast	**305**	**230**	**15.3**
Total or Avg.	**2,810**	**2,344**	**12.4**

VALERO

Demand on Limited Water Resources is Growing

So, How Much Water is on Earth? Here is How We Think of Water on Earth

5

VALERO

Demand on Limited Water Resources is Growing

Here is a Different View of the Water on Earth?

- The blue sphere sitting on the United States, reaching from Salt Lake City, UT to Topeka, KS has a diameter of about 860 miles, with a volume of about 332,500,000 cubic miles

6

Demand on Limited Water Resources is Growing

How Much Water is this in Context of our Solar System?

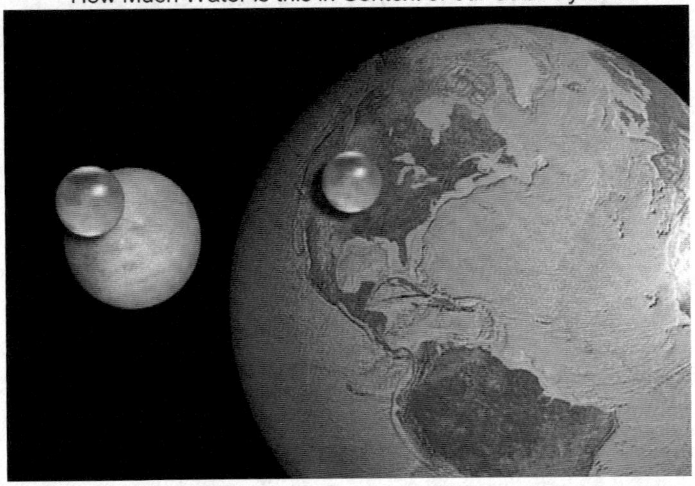

- This is a comparison of the water on Jupiter's moon Europa vs. the water on Earth – a very striking comparison.

7

Distribution of Earth's Water

Distribution of Earth's Water

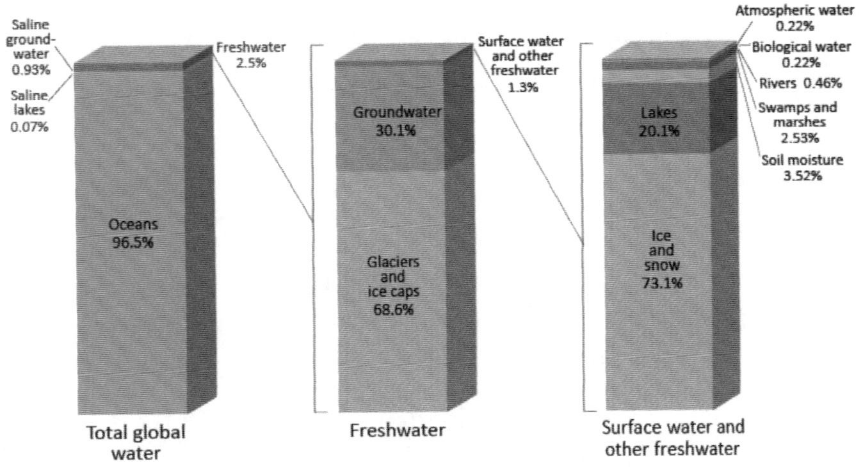

Source: Igor Shiklomanov's chapter "World fresh water resources" in Peter H. Gleick (editor), 1993, Water in Crisis: A Guide to the World's Fresh Water Resources.

8

Texas Water Use Survey Estimates

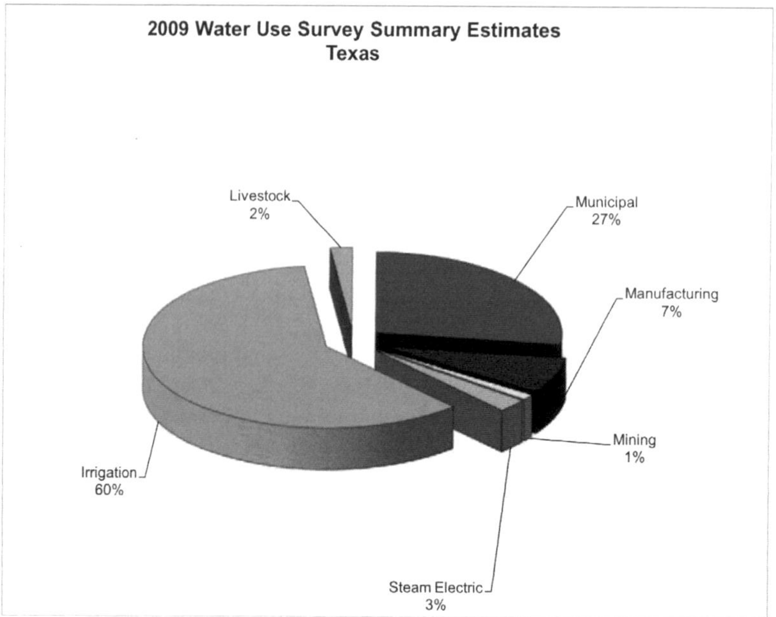

9

Refinery Water Use

- Depending on size and complexity, refinery water use varies from 1,400 gpm to 12,000 gpm
  - Majority of water from surface water supplies (reservoirs and rivers)
  - About a third of our refineries purchase clarified water from municipalities
  - Two refineries obtain water from wells that we own and operate
- Water efficiency varies from 0.4 bbl water per bbl of crude to 1.2 bbl of water per bbl of crude
  - Depends on refinery complexity and on-site hydrogen generation from steam methane reformers (water required for Rx)
  - Refineries in water rich areas consume more due to more use of steam drivers and cooling water circuits
  - Refineries in arid regions consume less with more electric drives and use of air fin fan cooling
- 30%-50% of the water is treated to make BFW make up with the remainder used for cooling tower make up and utility/wash down needs

Refinery Water Reuse and Conservation

- Good, detailed water balance is critical to any water conservation strategy
- Typical water reuse and conservation in place today includes:
  - Condensate recovery
  - Stripped sour water used for desalter make up and process wash water
  - RO reject for scrubber or fire water system make up
  - Waste water effluent for fire water system make up
  - Maximize cooling tower cycles
  - Blowdown control on boilers and cooling towers
- Most significant water loss is from evaporative cooling towers
- Typically 50%-65% of water supply is discharged as treated effluent

Other Water Reuse and Conservation Opportunities

- Opportunities exist to reduce cooling water needs and therefore reduce evaporative losses
  - Replace condensing turbines with electric motor drivers (currently implementing projects at several sites based on energy efficiency benefit)
  - Replace cooling water heat exchangers with fan fan coolers ($$$)
- Improving BFW treatment efficiency is another option
  - 2nd pass on RO reject to recover additional BFW make up
  - Install RO ahead of demin units to reduce regeneration cycles (in place at several refineries)
- Other opportunities are available but are often not cost effective:
  - Replace vacuum steam jets with electric vacuum pumps ($$$)
  - Use effluent for seal water ($$$ piping costs)
  - Tank farm and tracing condensate recovery ($$$ piping costs)

Alternative Water Supplies

- Treat and reuse effluent
  - At best, you could recover 60%-70% of effluent and return as raw water to the refinery or for cooling tower make up
  - Potentially provide 30%-40% of the total refinery water supply
- For coastal refineries, desalination is an option
- For refineries in urban areas, use of municipal treatment plant effluent is an option
  - Currently done at several US refineries

Water Reuse Challenges

- Economic projects are difficult to justify as long as current supply is available
- Typical cost to purchase and/or make clarified water is $1.50-$3.00 per 1,000 gallons
  - In arid regions, this cost may be as high as $6.00 per 1,000 gallons
- The cost to provide clarified water is typically less than 10% of the purchased energy cost to operate the refinery
- Recent evaluation completed at an arid refinery location (very low water usage) identified 14 potential conservation opportunities
  - Based on 10 year amortization, the recovered water cost ranged from $2 - $300 per 1,000 gallons
  - 4 of the projects were estimated at < $6/Mgal (cost of raw water supply) and are being pursued
    - + 175 gpm of water savings represents 6% of total water demand

Alternative Water Supply Challenges

- Use of treated municipal effluent water is feasible if piping costs can be minimized (close location)
- Cost to own and operate either a coastal desalination facility or effluent recovery system sized for a single refinery is estimated at \$8.00 - \$12.00 per 1,000 gallons
  - Estimate assumes brine may be discharged, deep well injection not required
- To obtain economies of scale, an industrial community based facility is needed
- Who wants to be first in?

Conclusions

- With continued growth, competition will increase for limited fresh water supplies
- Optimizing water usage with operational and low capital improvements can reduce consumption by 5-15%
- Effluent treatment/reuse and desalination of seawater are both very expensive (capital and operating expense)
- An industry-wide approach to optimize water across all the industrial water users is needed
  - Optimize how money is spent to reduce water demands to get the biggest bang for the buck
  - Can't afford to proactively make the necessary large capital and operating investments if our competition is not doing the same
- The other concern is that eventually a water supply strategy may need to be included as part of the project process for a new unit, plant or expansion
  - So, here also, early movement on reducing water use could put you at a disadvantage

Water Reuse
in the Petroleum Industry

Jim Myers, P.E.
Chevron

Why Water Reuse?

Thirsty Planet

- World-wide Water Shortages

Save Money/Save Energy/Sustainability

- Produce More Water Than Oil
- Water Treatment Influences Profitability
- Water Companies in Denial

Do More Than Just Check the Environmental Social Impact Assessment (ESIA) Boxes

- Environmental Stewardship
- Community Outreach
- Environmental Education
- Sustainability

2

Water Reuse Outline

Introduction to Water Reuse
- Water Sources
- Treatment
- Reuse Standards

Case Examples
- Applications
- Upstream
- Downstream

Future of Water Reuse
- Ideas to Go
- Barriers and Fixes

3

Water Reuse Opportunity

Brief History of Petroleum Industry
- Kerosene replaces whale oil
- Light bulb replaces kerosene
- Fuel oil replaces coal in ships
- Crude refined into gasoline and lube oils
- Fertilizer and pharmaceuticals
- Water usually increases with age of field
- 0 to >100 bbl of water for 1 bbl of oil
- 1.3 million tonnes produced water per tonne of hydrocarbons produced

Human Population
- 6.6 billion September 2007
- 9 billion 2050
- 40% short on water by 2050

World-wide Water Shortage
- 70% earth is covered with water
- 97% saltwater
- 3% freshwater
- <1% freshwater is readily available
- Scarcity could limit growth

Opportunity for Produced Water Reuse
- Typically produced water is found water not considered a water source
- Convert by-product into an asset
- Reuse water for agriculture
- Grow biofuels
- Sequester carbon

Convergence of Petroleum Industry, Water Shortage and Human Population

4

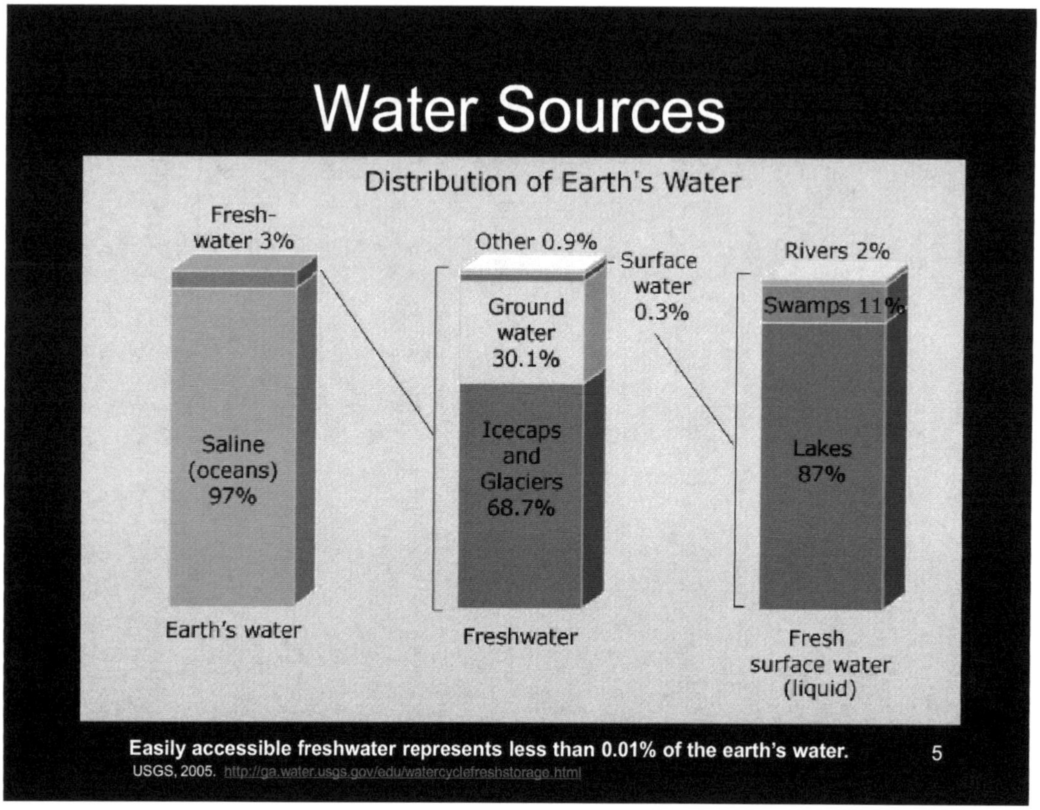
Water Sources
Distribution of Earth's Water
Fresh-
water 3%
Saline
(oceans)
97%
Earth's water
Other 0.9%
Ground
water
30.1%
Icecaps
and
Glaciers
68.7%
Freshwater
Surface
water
0.3%
Rivers 2%
Swamps 11%
Lakes
87%
Fresh
surface water
(liquid)
Easily accessible freshwater represents less than 0.01% of the earth's water.
USGS, 2005. http://ga.water.usgs.gov/edu/watercyclefreshstorage.html
5

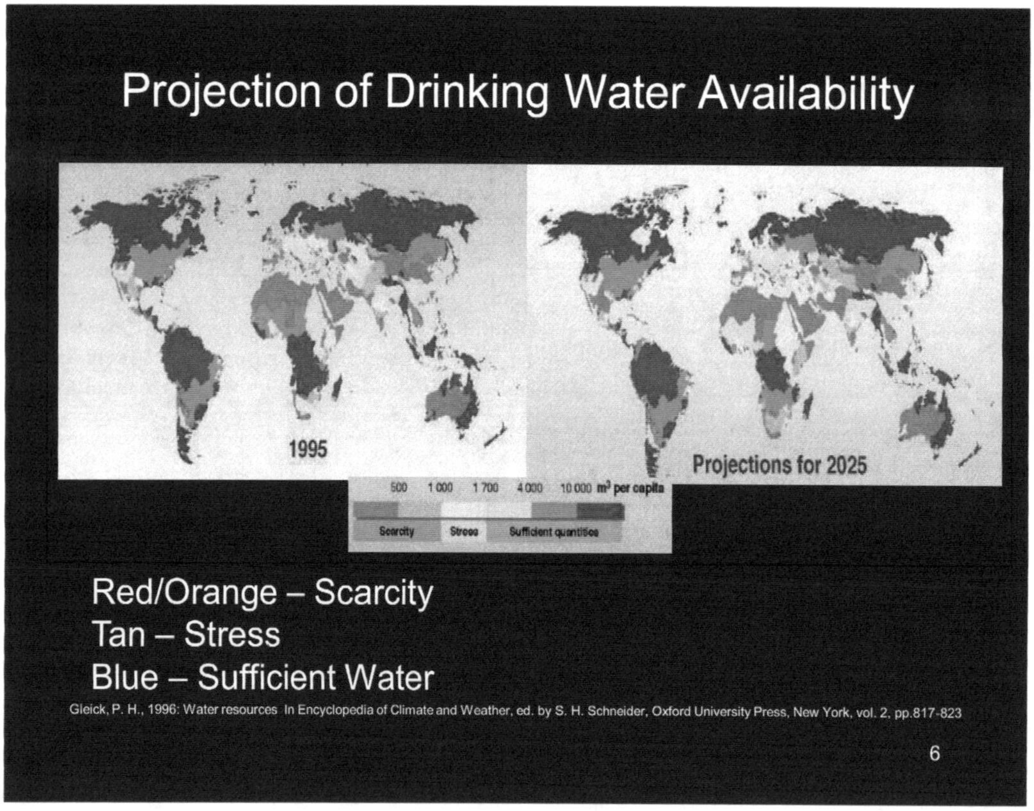
Projection of Drinking Water Availability
1995
Projections for 2025
500 1 000 1 700 4 000 10 000 m³ per capita
Scarcity
Stress
Sufficient quantities
Red/Orange – Scarcity
Tan – Stress
Blue – Sufficient Water
Gleick, P. H., 1996: Water resources In Encyclopedia of Climate and Weather, ed. by S. H. Schneider, Oxford University Press, New York, vol. 2. pp.817-823
6

Water Use and Cost

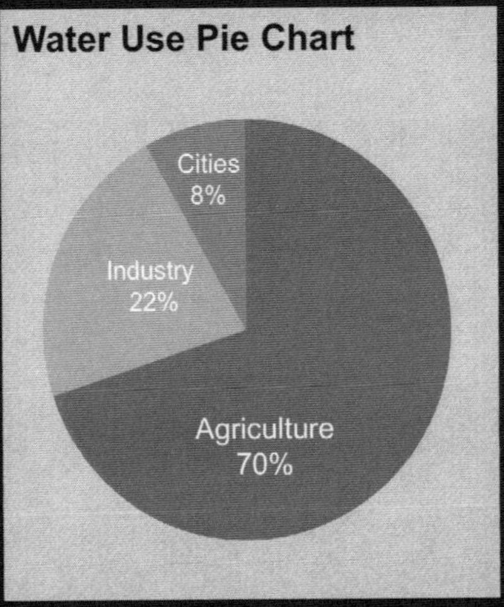

Whyte, Fergal, 2006, "Cities of the Future – Urban Sustainability & Water"

Water Source	Unit Cost	Cost ($/acre-ft)
Irrigation Water		
SE Texas		$10 to 100
California (wet year)		$100 to 150
California (dry year)		$200 to 500
US Cities ($/1000 gal)		
Low (Mississippi)	$ 0.80	$ 261
Average	$ 2.49	$ 811
High (West Virginia)	$ 5.61	$ 1,828
Treated PW (1 bbl)	$ 0.01	$ 78
	$ 0.02	$ 155
	$ 0.03	$ 233
	$ 0.04	$ 310
	$ 0.05	$ 388
	$ 0.06	$ 465
	$ 0.07	$ 543
	$ 0.08	$ 621
	$ 0.09	$ 698
	$ 0.10	$ 776
	$ 0.20	$ 1,552
	$ 0.30	$ 2,327
	$ 0.40	$ 3,103
	$ 0.50	$ 3,879
	$ 1.00	$ 7,758
16 oz Bottled Water	$ 1.00	$ 2,606,630

Acre-ft ~ 209 x 209 x 1 ft ~ 325,829 gal

7

Human Population and Water

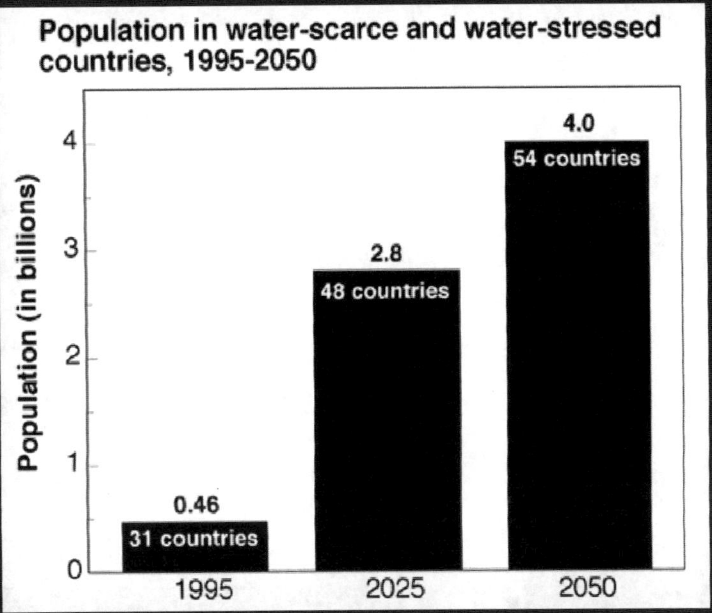

AAAS Atlas of Population and Environment, 1998. http://www.ourplanet.com/aaas/pages/natural03.html#

8

Water Reuse PFD

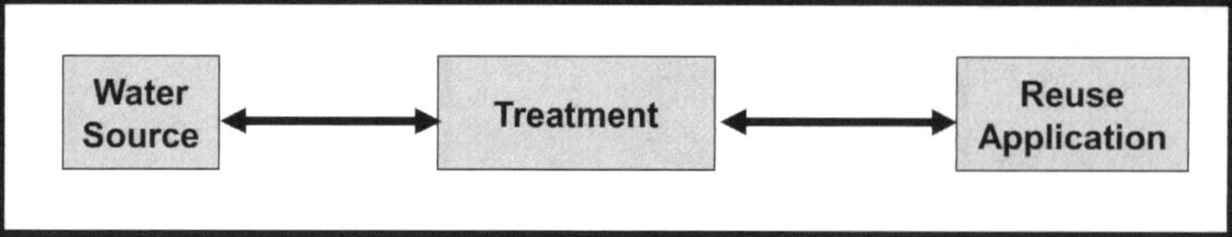

- Simple as A-B-C, 1-2-3 or "1st name, middle name, last name"
  - Lots of Water Sources
    - Produced Water, Refinery Wastewater, Marketing Terminal Water, Ground Water, Storm Water, Parking Lot Runoff
  - Plenty of off-the-shelf Treatments
  - Unlimited Supply of Applications

PFD – Process Flow Diagram

Typical Water Treatment PFD

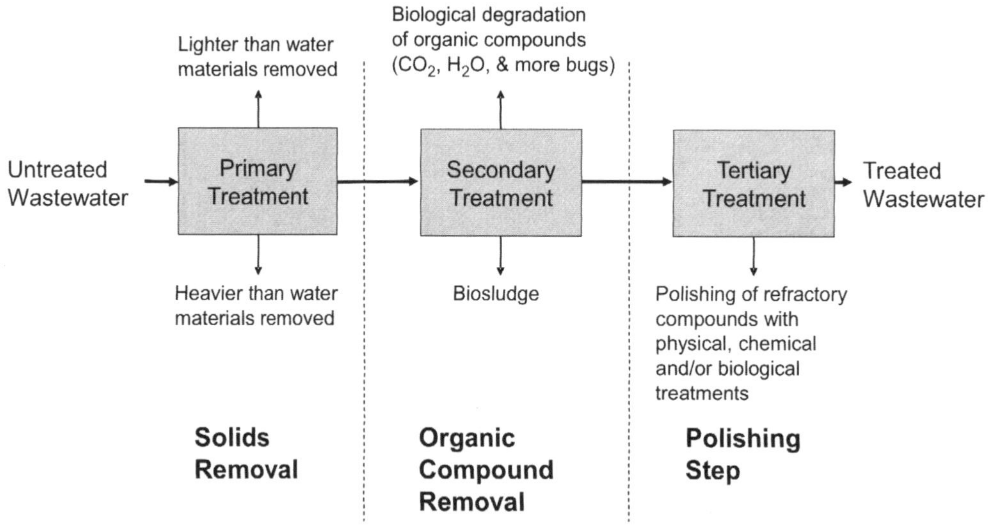

PFD – Process Flow Diagram

Typical Produced Water Treatment PFD

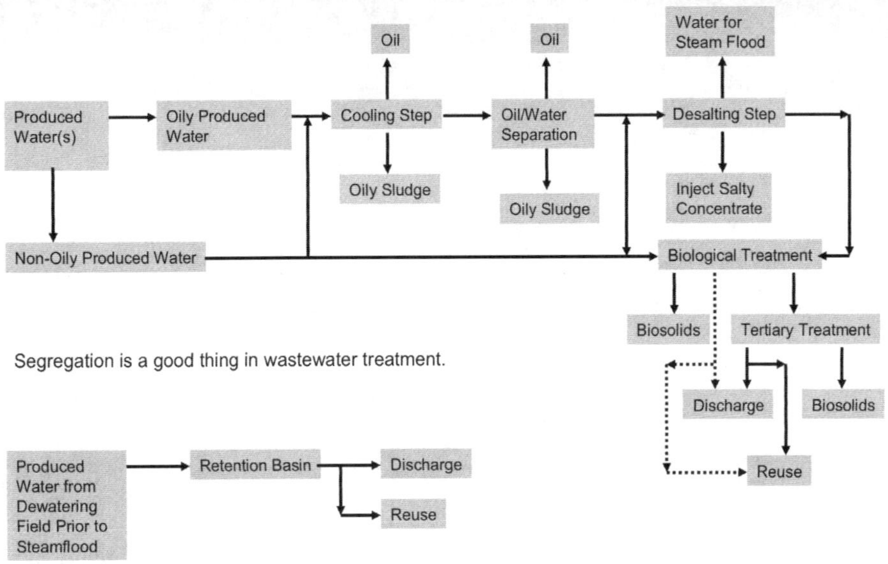

1st Step in Design is to Create a Series of PFDs.

PFD – Process Flow Diagram

11

Water Treatment Technologies

Wide Selection – One Size Does Not Fit All

Water Treatment Grouped by Function	Comment
Primary Treatment - Oil/Water Separator - API* - CPI* - PPI* - DAF/DNF* - IAF* - Clarifier - Hydrocyclone - Treatment Wetland	Gravity settling of solids heavy than water. Float oils lighter than water. Entrain oil with air/gas bubbles. Use density differences to remove solids and oils.
Secondary Treatment - ASTU* - SBR* - Facultative Lagoon - RBC - Aerated Lagoon - Anaerobic Filter - Treatment Wetland - MBR* - Trickling Filter - GAC*	Biological treatment uses micro-organisms to consume organic compounds in an ambient temperature combustion reaction. Some of the treatments are energy intensive with small physical footprints. Others are not energy intensive but have large physical footprints.
Tertiary Treatment - Activated Carbon - Filtration - Chlorination - UV* - Chemicals (FeCl, O3, H2O2)* - Treatment Wetland	Polishing treatments remove the refractory compounds by physical, chemical and biological means.
Salts (desalting) - RO* - FO* - ED* - NF* - Freezing - UF*	Salt removal is typically to meet irrigation water standards (EC/TDS/salinity). Some of the membrane technologies require pre-treatment (solids removal) to avoid membrane fouling.

*SEE ACRONYM LIST – LAST SLIDE

12

Water Reuse Applications

Find Out What Local Stakeholders Need

Category	Typical Application
Agriculture	Crop irrigation, Commercial nurseries
Livestock	Cattle, Sheep, Pigs, Chickens, Waterfowl
Aquaculture	Fish, Shrimp, Frogs, Bait
Silviculture	Tree farming
Landscape Irrigation	Residences, Golf courses, Parks, Cemeteries, Freeway medians, Greenbelts, School yards
Industrial Recycling and Reuse	Boiler feed water, Fire protection, Cooling tower, Process water, Manufacturing, Vehicle washing, Dust control
Groundwater Recharge	Groundwater replenishment (water banking), Salt water intrusion control, Subsidence control
Recreation	Lakes, streams and ponds, Snow making
Wildlife Habitat	Marsh restoration, Habitat enhancement, Fisheries
Nonpotable Urban Use	Fire protection, Air conditioning, Toilet flushing
Potable Use	Blending in water supply reservoirs, Blending in groundwater, Direct pipe to pipe water supply

13

Brief Overview on Water Reuse Standards

Pick and Choose to Match Application

Reference	Category/Comment
Health Guidelines for Use of Wastewater for Agriculture and Aquaculture by WHO (1989)	Focus on helminth removal
Guidelines for Water Reuse by USEPA (1992)	Focus on municipal water
Water Reuse[13] (2007)	Compilation of municipal and agricultural standards
Water Encyclopedia[14] (2007)	Compilation of water standards
USA State Regulations on Water Reuse	Most have USEPA like standards
Forestry Standards (several sources & university papers)	Trees tend to tolerate lower quality water than other plants
Fishery Standards (several sources & university papers)	Focus is the taste of fish in commercial fish farming operations
AB (Alberta) Canada	Comprehensive lists including treatment technologies

WHO – World Health Organization USEPA – United States Environmental Protection Agency

14

Water Reuse Analytical Testing Parameters for Irrigation – Salt Focus

		Degree of Restriction on Use		
Irrigation parameter	**units**	**None**	**Slight to Moderate**	**Severe**
		Salinity		
EC	**dS/m**	**<0.7**	**0.7-3.0**	**>3.0**
TDS	**mg/L**	**<450**	**450-2000**	**>2000**
Sodicity				
SAR, 0-3		**and EC ≥0.7**	**0.7-0.2**	**<0.2**
3-6		**≥1.2**	**1.2-0.3**	**<0.3**
6-12		**≥1.9**	**1.9-0.5**	**<0.5**
12-20		**≥2.9**	**2.9-1.3**	**<1.3**
20-40		**≥5.0**	**5.0-2.9**	**<2.9**
Specific Ion Toxicity				
Sodium (Na)	**SAR**	**<3**	**3-9**	**>9**
Surface irrigation	**mg/L**	**<70**	**>70**	**>350**
Sprinkler irrigation	**mg/L**	**<140**	**140-350**	**>3.0**
Chloride (Cl)	**mg/L**	**<100**	**>100**	
Surface irrigation	**mg/L**	**<0.7**	**0.7-3.0**	
Sprinkler irrigation				
Boron (B)				
Miscellaneous effects				
Nitrogen (total N)	**mg/L**	**<5**	**5-30**	**>30**
Bicarbonate (HCO_3) (overhead sprinkling only)	**mg/L**	**<90**	**90-600**	**>500**
pH	**s.u.**		**Normal range 6.5-8.4**	
Residual chlorine (overhead sprinkling only)	**mg/L**	**<1.0**	**1.0-5.0**	**>5.0**

EC – Electrical Conductivity TDS – Total Dissolved Solids SAR – Sodium Adsorption Ratio

15

Petroleum Industry Water Reuse Case Examples

Case	Treatment	Reuse	Comment
California Produced Water[3]	Oil/Water Separation, Walnut Shell Filter, Cooling Pond	Agriculture	Chemical and mechanical treatment. Small footprint. Reused water grows >40 crops.
California Produced Water	Oil/Water Separation, Cooling, RO, CTW	Agriculture	RO for salt removal to meet irrigation water standards. Reused water recharges irrigation aquifer.
California Produced Water Demonstration Wetland	Oil/Water Separation, Walnut Shell Filter, Cooling Pond, Treatment Wetlands	Agriculture	Test program capturing data from multiple cells, plant species, planting densities and construction costs.
Ohio Closed Refinery Storm & Ground Water	Oil/Water Separation, GAC, Pond, CTW	Wildlife	Wetland is winning wildlife awards (pollinator and waterfowl). WHC certified wildlife management plan.
Kansas Refinery Wastewater	Oil/Water Separation, ASTU, Facultative Lagoon, CTW	Wildlife	Gravity fed CTW. Pump once. Let gravity do the rest. Large wetland is attracting wildlife use. IOGCC award. WHC certified.
Wyoming DOE CRADAs Pilot Studies	Oil/Water Separation, Cooling, Facultative Lagoon, CTW	Wildlife Agriculture Aquaculture	Series of pilot CTW studies. Reused water irrigates a normally dry stream creating grazing habitat for mule deer and elk.
Far East Produced Water Feasibility Study	Oil/Water Separation, Cooling Pond, CTW	Agriculture Aquaculture Silviculture	Several applications possible. Lack of infrastructure makes water hand-off challenging.
Africa Produced Water Feasibility and Pilot Studies	Oil/water Separation, Cooling Pond, CTW	Agriculture Livestock	Several applications possible. Lack of infrastructure makes water hand-off challenging.

16

RO – Reverse Osmosis CTW – Constructed Treatment Wetland GAC – Granular Activated Carbon ASTU – Activated Sludge Treatment Unit
DOE – Department of Energy CRADA – Cooperative Research and Development Agreement

Water Reuse Case Example
Refinery WWTP PFD

Refinery Wastewater → Primary Treatment → Secondary Treatment → Tertiary Treatment → Treated Water

FORM	Oil/Water Separation Solids Removal	Activated Sludge Facultative Lagoon	**Treatment Wetland**
FUNCTION	Removal of Oil and Solids	Removal of Water Soluble Organic Compounds (WSO)	**Removal of WSO and Total Suspended Solids (TSS)**
OBJECTIVE	Setup for Biological Treatment	Meet Discharge Limits	**Meet Discharge Limits** **Provide Wildlife Habitat**

WWTP – Wastewater Treatment Plant PFD – Process Flow Diagram
WSO – Water Soluble Organic Compounds TSS – Total Suspended Solids

17

Water Reuse Case Example
Refinery Wastewater Treatment Wetland

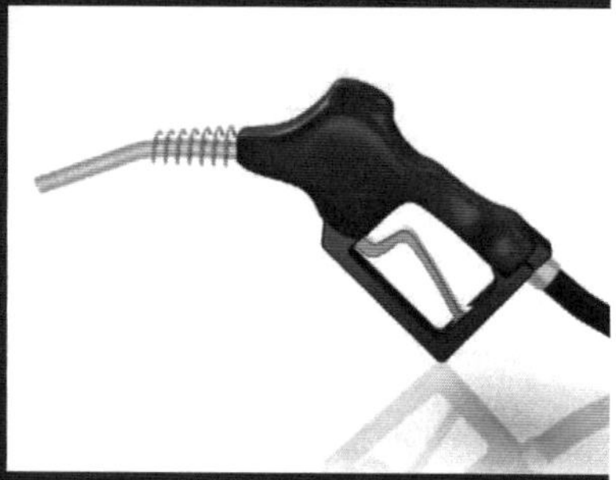

- 90-acre Site in Mid-West
- Former Soybean Field
- 2 mgd Average Flow
- Multiple Cells
- Parallel or Series Flow
- Wastewater and Wildlife Balance
  - W-ditches, 6:1 & 4:1 Levees, Gravel/Grit, Hedgerow/Windbreak, Nesting Boxes, Biodiversity Increase (60 to 250 species)
- Interstate Oil & Gas Compact Commission (IOGCC) and Wildlife Habitat Council (WHC) Award

18

Refinery Water Reuse Example

Aerial View of Treatment Wetland

19

Refinery Water Reuse Example

Mallard Landing on Cell Nos. 4 and 5

20

Water Reuse Case Example Department of Energy (DOE) Produced Water Wyoming

- 3 in Operation in the Western US
- Economics Good
- Wildlife Benefits Better
- Discharge Limits Met

21

DOE Wyoming Produced Water Reuse Case Example

Produced Water WWTP PFD

Produced Water → Primary Treatment → Secondary Treatment → Tertiary Treatment → Treated Water

FORM	Cooling Tower Oil/Water Separation Solids Removal	Facultative Lagoon	**Treatment Wetland**
FUNCTION	Water Cooling Removal of Oil and Solids	Water Cooling	**Removal of WSO**
OBJECTIVE	Meet Discharge Limits Exclude Wildlife (nets)	Setup for Wetland	**Meet Discharge Limits** **Water Reuse for Wildlife**

WWTP – Wastewater Treatment Plant PFD – Process Flow Diagram

22

DOE Wyoming Produced Water Reuse
Beneficial Reuse for Wildlife

23

Water Reuse Ideas to Go
Water Reuse Justification Kit

Economics

- CAPEX
- OPEX

GHG Emission Reduction

- CO2, NOx, SOx, CO, PM, Energy Usage
- Soil Carbon Sequestration

ESIA

- Environmental
- Social
- Health

Business Development

- Political
- Regulatory
- Zero Water Footprint

CAPEX – Capital Expense OPEX – Operational Expense GHG – Greenhouse Gases ESIA – Environmental Social Impact Assessment
CO2 – Carbon Dioxide NOx – Nitrous Oxides SOx – Sulfur Oxides PM – Particulate Matter

24

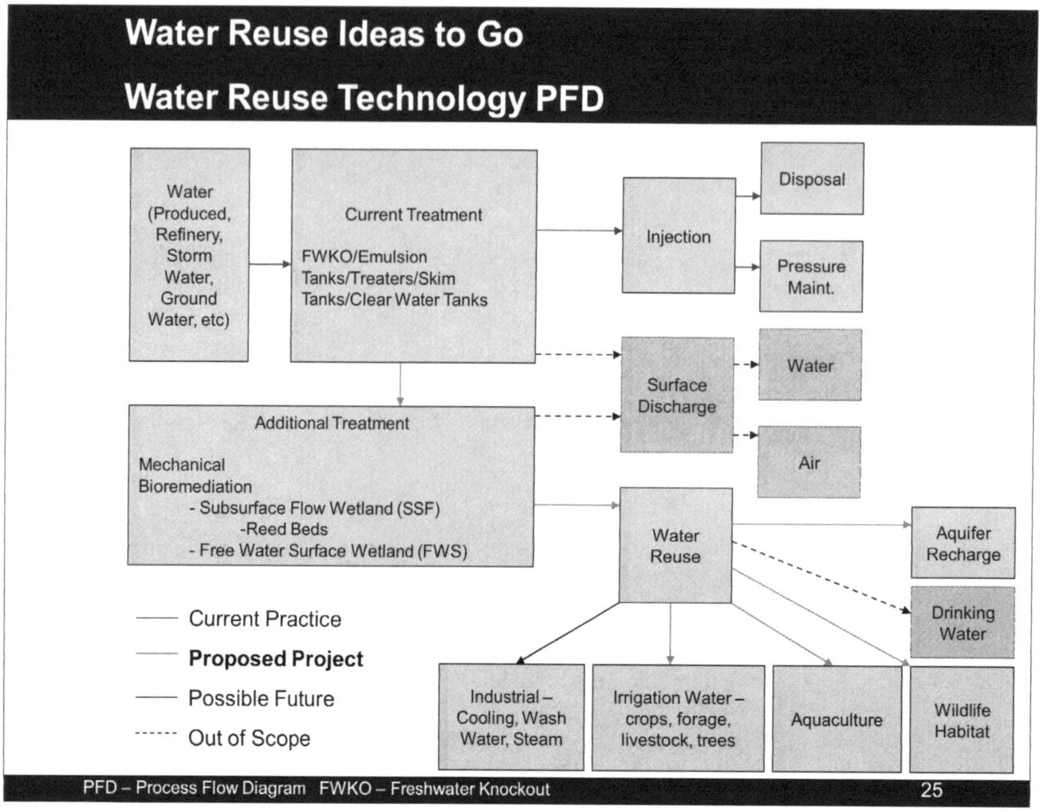
Water Reuse Ideas to Go
Water Reuse Technology PFD
Water (Produced, Refinery, Storm Water, Ground Water, etc)
Current Treatment
FWKO/Emulsion Tanks/Treaters/Skim Tanks/Clear Water Tanks
Injection
Disposal
Pressure Maint.
Surface Discharge
Water
Air
Additional Treatment
Mechanical
Bioremediation
- Subsurface Flow Wetland (SSF)
-Reed Beds
- Free Water Surface Wetland (FWS)
Water Reuse
Aquifer Recharge
Drinking Water
Current Practice
Proposed Project
Possible Future
Out of Scope
Industrial – Cooling, Wash Water, Steam
Irrigation Water – crops, forage, livestock, trees
Aquaculture
Wildlife Habitat
PFD – Process Flow Diagram FWKO – Freshwater Knockout
25

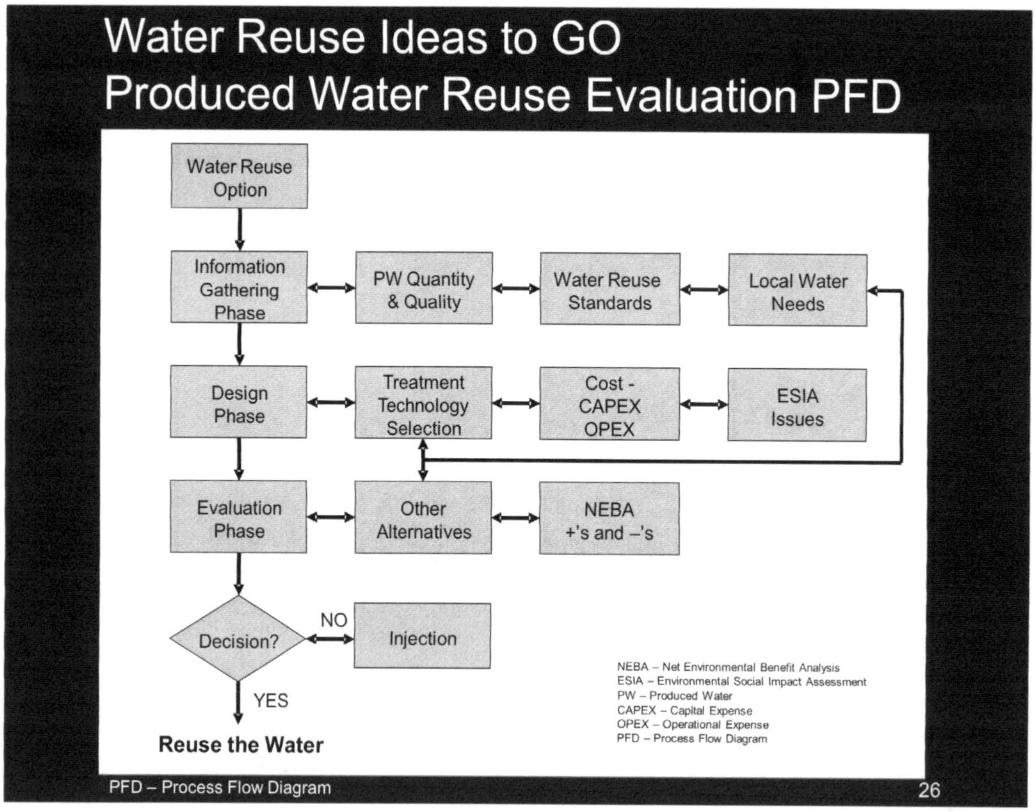
Water Reuse Ideas to GO
Produced Water Reuse Evaluation PFD
Water Reuse Option
Information Gathering Phase
PW Quantity & Quality
Water Reuse Standards
Local Water Needs
Design Phase
Treatment Technology Selection
Cost - CAPEX OPEX
ESIA Issues
Evaluation Phase
Other Alternatives
NEBA +'s and –'s
Decision?
NO
Injection
YES
Reuse the Water
NEBA – Net Environmental Benefit Analysis
ESIA – Environmental Social Impact Assessment
PW – Produced Water
CAPEX – Capital Expense
OPEX – Operational Expense
PFD – Process Flow Diagram
PFD – Process Flow Diagram
26

Water Reuse Barriers

Barriers and Fixes

- **Barrier – risk aversion and "we have never done it that way before"**
- Fix – "Ask them what their 2nd answer is", education, economic evaluations, pilot studies, risk assessments and ESIA.
- **Barrier -1st to be 2nd to use a new technology mind set**
- Fix – Repackage off-the-shelf technologies and use water reuse best management practices (BMPs).
- **Barrier – Regulations do not encourage Water Reuse**
- Fix – Help change the regulations to benefit local stakeholders (farmers, ranchers, municipalities, etc).
- **Barrier – TMI overload and analysis paralysis**
- Fix – KISS principle (water source → treat → reuse)

27

ESIA – Environmental Social Impact Assessment BMPs – Best Management Practices TMI – Too Much Information KISS – Keep It Simple Stupid

Water Reuse Future

Water Reuse Train Has Left the Station

- **Voluntary – set your own terms**
- **Involuntary – do what you are told**

Regulations Encouraging Water Reuse

- **Incentives and Penalties**

Compilation of Water Reuse Standards

Regional Water Banks for Trading Water

Beg, Borrow and Steal BMPs from Existing Water Reuse Projects

Corporate Metrics Reporting Water Reuse

28

BMPs – Best Management Practices

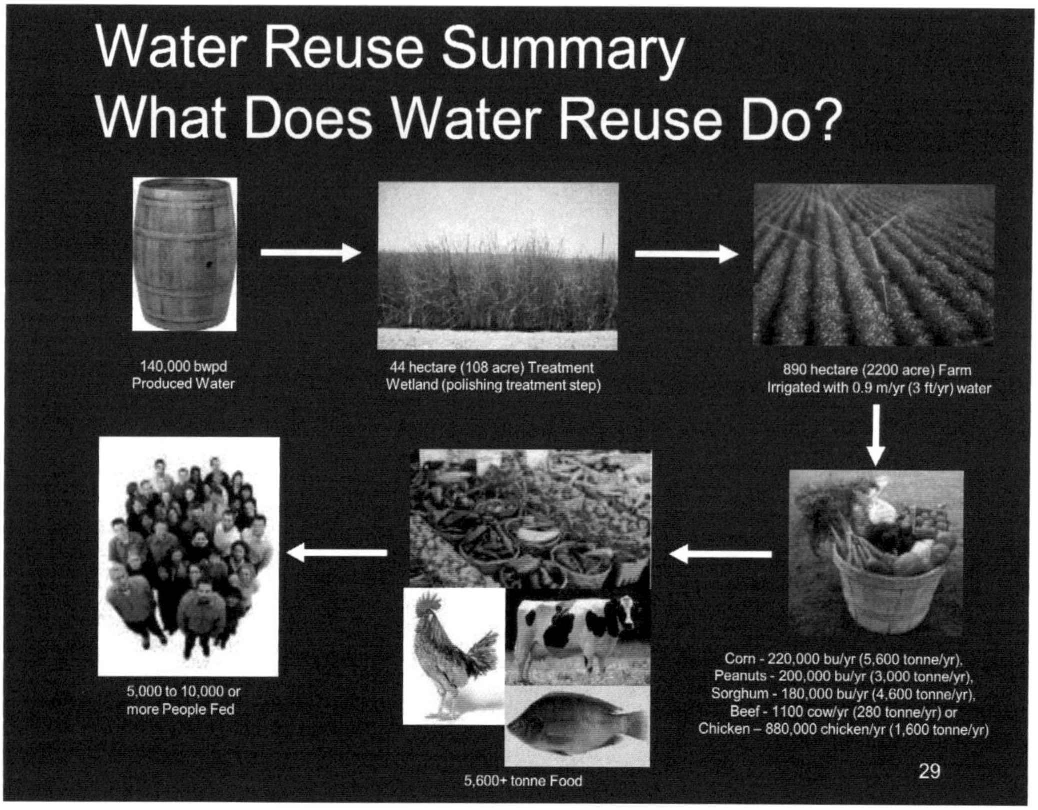

Acronym List

ASTU – ACTIVATED SLUDGE TREATMENT UNIT
API – AMERICAN PETROLEUM INSTITUTE
CPI – CORRUGATED PLATE INTERCEPTOR
PPI – PARALELL PLATE INTERCEPTOR
DAF – DISSOVLED AIR FLOTATION
DNF – DISSOLVED NITROGEN FLOTATION
IAF – INDUCED AIR FLOTATION
RBC – ROTATING BIOLOGICAL CONTACTOR
RO – REVERSE OSMOSIS
FO – FORWARD OSMOSIS
NF – NANO-FILTRATION
UF – ULTRA-FILTRATION
ED – ELECTRODIALYSIS
SBR – SEQUENCING BATCH REACTOR
PFD – PROCESS FLOW DIAGRAM
MBR – MEMBRANE REACTOR
GAC – GRANULAR ACTIVATED CARBON
UV – ULTRAVIOLET
O3 – OZONE
FeCl – FERRIC CHLORIDE
H2O2 – HYDROGEN PEROXIDE
WWTP – WASTEWATER TREATMENT PLANT
EC – ELECTRICAL CONDUCTIVITY
TDS – TOTAL DISSOLVED SOLIDS
ESIA – ENVIRONMENTAL SOCIAL IMPACT ASSESSMENT
GHG – GREENHOUSE GASES
OPEX – OPERATIONAL EXPENSE
CAPEX – CAPITAL EXPENSE
USEPA – UNITED STATES ENVIONMENTAL PROTOTECITON AGENCY
SAR – SODIUM ADSORPTION RATIO
WHO – WORLD HEALTH ORGANIZAITON
CO2 – CARBON DIOXIDE
NOx - NIRTOUS OXIDES
SOx – SULFUR OXIDES
PM – PARTICULATE MATTER
BMP – BEST MANAGEMENT PRACTICE
TMI – TOO MUCH INFORMATION
KISS – KEEP IT SIMPLE STUPID
CTW – CONSTRUCTED TREATMENT WETLAND
CRADA – COOPERATIVE RESEACH AND DEVELOPMENT AGREEMENT
IOGCC – INTERSTATE OIL & GAS COMPACT COMMISSION
WHC – WILDLIFE HABITAT COUNCIL
WSO – WATER SOLUBLE ORGANIC COMPOUNDS
PW – PRODUCED WATER
NEBA – NET ENVIRONMENTAL BENEFIT ANAYLYSIS

GE Power & Water
Water & Process Technologies

AIChE Industrial Water Reuse

Applying Advanced Technologies to Reuse Applications

Erik Hanson
Director of Product Management, Systems

imagination at work

May 1, 2013

GE's Portfolio

Our leadership in equipment solutions:

- Advanced ultrafiltration, membrane bioreactor, reverse osmosis membranes and membrane chemistries
- Mobile fleet and water outsourcing capabilities
- Tough-to-treat applications, such as unconventional fuels and mining
- Packaged water treatment equipment
- Analytical instruments for measuring water quality

Our leadership in chemical and knowledge management solutions:

- Cooling and boiler water technologies that enable customers to protect their assets
- Chemical treatment for ethylene, styrene and elastomer production facilities
- Refinery treatment solutions focused on tough-to-treat crudes
- Remote monitoring and diagnostic solutions

imagination at work

US Power Plant

Reusing Municipal Waste Effluent

imagination at work

RePAK* UF-RO at US Power Plant

Challenge: Lower cost of make-up cooling water in a northeast US Power Plant
Solution: RePAK UF-RO treats municipal waste effluent to cooling water quality

Trash Burning Power Plant in Pennsylvania

- Initially using city water to feel 900 gpm make-up cooling water
- Rate increases hurting profitability of power plant
- Negotiated low cost long term off-take agreement for effluent with nearby Waste water treatment plant
- Scope: Two GE RePAK UF-ROs
- Approximate equipment cost of $1.0MM

imagination at work

Process Flow

Feed Water Quality Limits (UF feed)

- Ave. TSS / TOC < 15 mg/L Peak =< 50 mg/L (72hrs)
- BOD Ave. < 15 mg/L Peak =< 30 mg/L (72 hrs)
- COD Ave. < 45 mg/L Peak =< 100 mg/L (72 hrs)
- TDS = < 3000 mg/L. Temperature = 15-30C

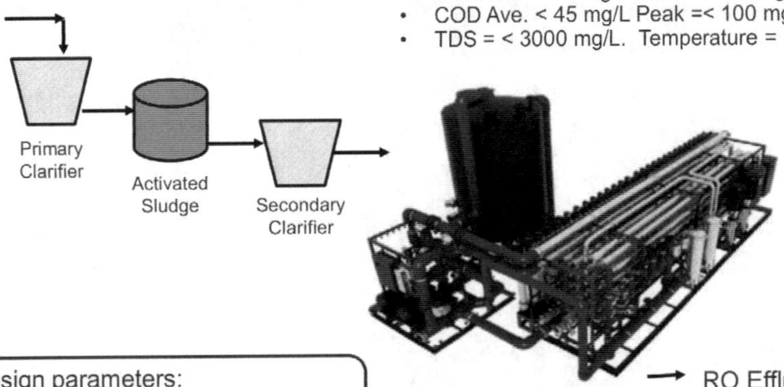

Design parameters:

- UF Average Flux: 30 gfd
- RO Average Flux: 11-12 gfd

RePAK - UF+RO Tertiary Package

RO Effluent:

- ~ 95% reduction in TDS
- < 1mg/L BOD
- < 10 mg/L COD

Re-PAK Scope Outline

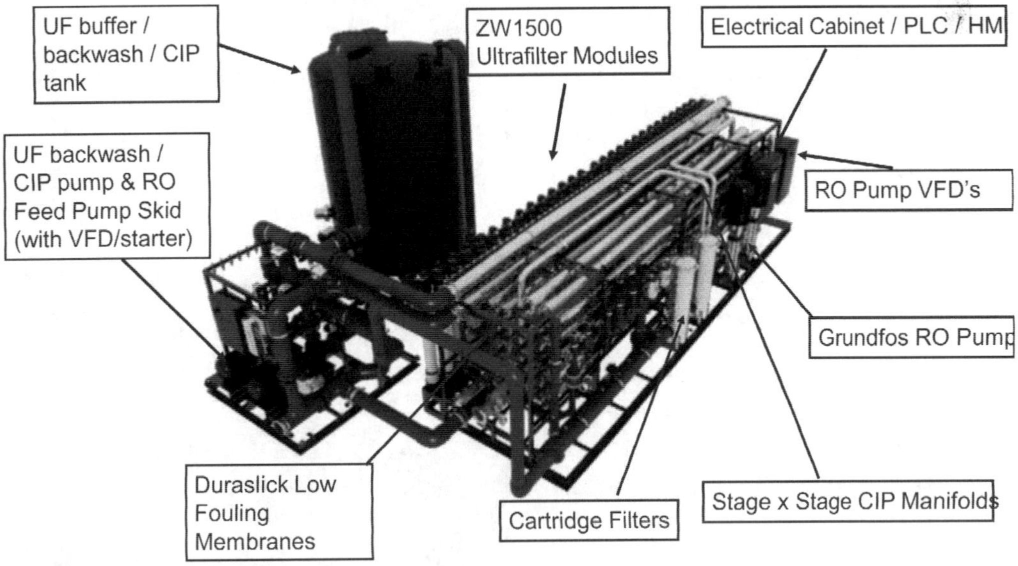

Not shown and shipped separately are:
UF feed pump, Self Cleaning Screen, Modulating Feed Valve (pressure), and Air Blower package for air wash

Coca Cola in Asia
Improving Water Yield

imagination at work

AquaSel* at Coca-Cola Plant: Near Zero Liquid Discharge

Challenge: Relieve water shortage & improve water yield in a beverage facility
Solution: AquaSel (non-thermal brine concentrator) with 36,000 gpd capacity

In 1000 hours of pilot operation:

- AquaSel captured and converted 1.5 million gallons (5,678 m3) of a water waste stream for reuse
- Captured water quality was equal to incoming fresh water source
- AquaSel system recovered 99%+ of ingredient room water waste stream

imagination at work

Innovative Technology to Save Billions of Gallons of Water Per Year.

Today

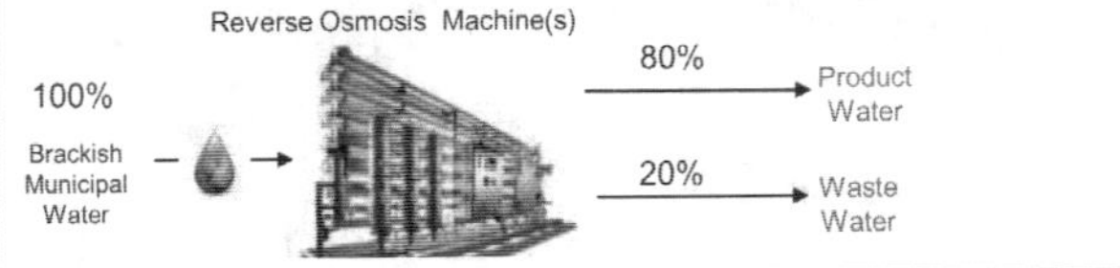

- 20% of purchased water goes to waste
- Over 1300+ Installed GE Systems

Tomorrow

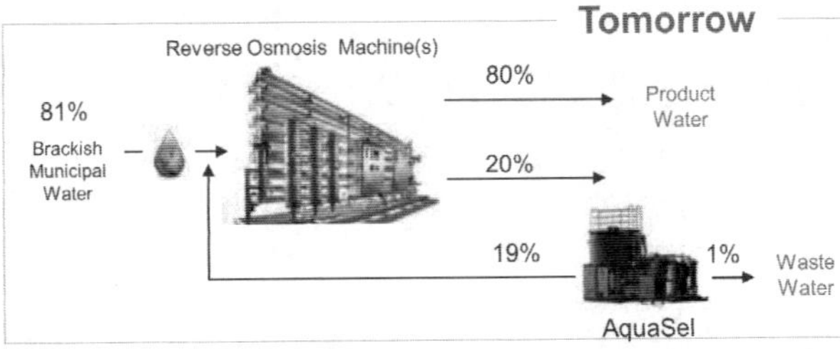

- Bolt on proprietary technology
- Waste water ↓90%
- Purchased water ↓ 10%
- Saving billions of gallons of water per year at Bottlers
- **Supports sustainability metrics**

Improving Water Use Ratio

ecomagination

imagination at work

AquaSel Flow Sheet

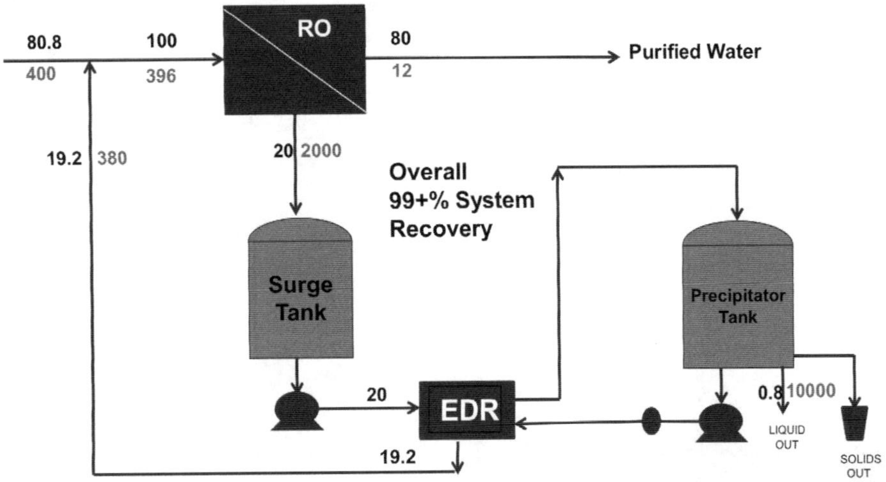

1. **Black Numbers are units of Flow**
2. **Red Numbers are units of TDS**

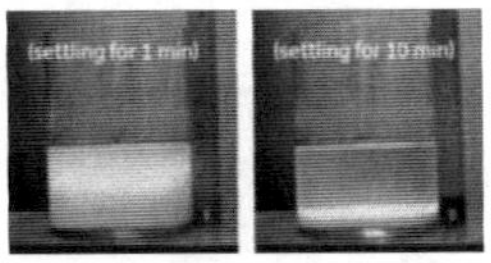

imagination at work

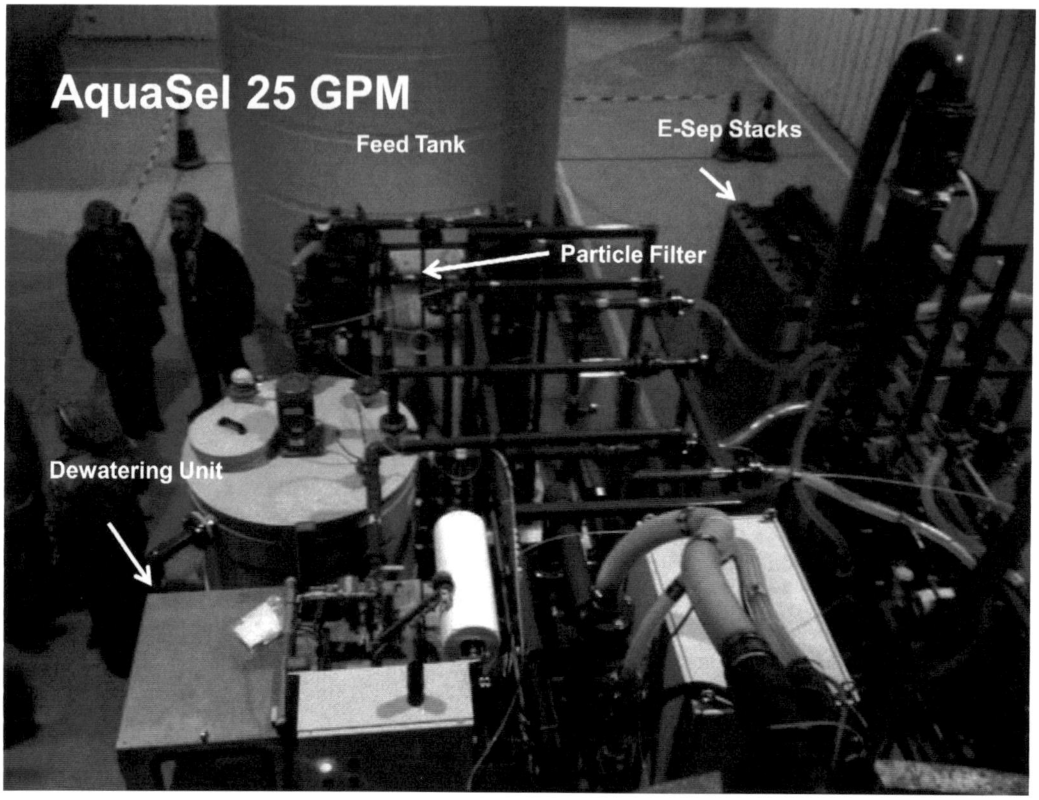

E-Separation Process Overview

Use electrical potential to drive ionic salts through the membranes

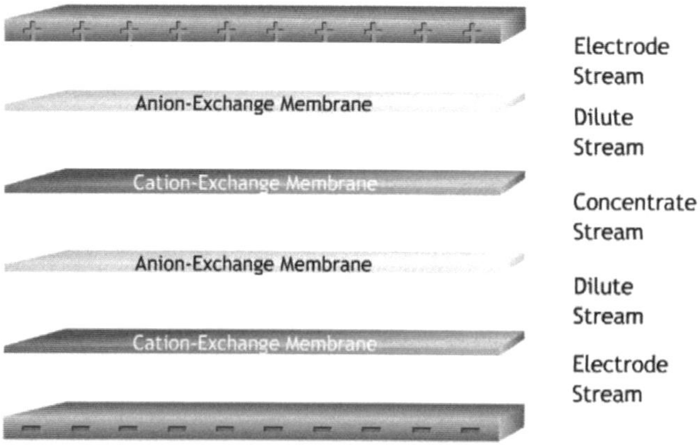

imagination at

Polarity Reversal

- Minimizes scaling and colloidal deposition
- Breaks up freshly precipitated scale and flushes them before they grow and cause damage
- Reduces slime or similar formations on membrane surfaces

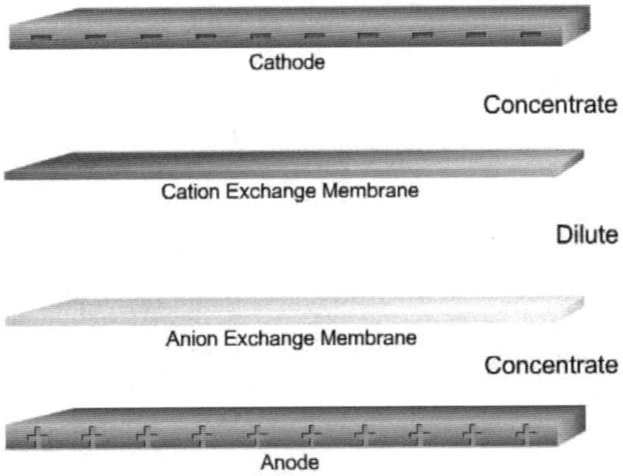

imagination at work

More Case Studies

imagination at work

Harvesting wastewater in Australia

Challenge: Ongoing drought challenged availability of water for golf course

Solution: Sewer mining water reuse plant provides irrigation water

Pennant Hills Golf Club, Australia's first commercial sewer mining water reuse plant

- Conserves 25 million gallons of Australia's fresh water a year
- Advanced MBR produces 172,000 gallons of high quality water per day which is used to irrigate 55 acres
- "We are proud to be the first to embrace this innovative approach. It is bringing us a drought-proof supply of water that minimizes impact on Australia's fresh water reserves." — Steve Walker, president, Pennant Hills Golf Club

imagination at work

17
Water & Process Technologies

Creating "NEWater" in Singapore

Challenge: Inadequate supplies of renewable fresh water

Solution: Treat and reuse wastewater effluent for local industry

Bedok NEWater Factory

- Transforming wastewater into high quality industrial feedwater and potable water
- The final product is termed "NEWater"
  - Initially used as a feed for the electronics industry, wafer fabrication plants, and commercial building cooling towers
- A growing percentage is released back into local reservoirs for indirect potable reuse applications

imagination at work

18
Water & Process Technologies

Thank You

imagination at work

Availability of Models to Estimate Greenhouse Gas Emissions & Carbon Footprint of Water Reuse Facilities

presented at:
AIChE Industrial Water Use and Reuse Workshop: April 30 - May 1, 2013

James R. Mihelcic, PhD, BCEEM
University of South Florida, Tampa
jm41@usf.edu

&

Qiong Zhang, PhD; Mark V. Santana; Pablo K. Cornejo;
Andrea Rocha, PhD; Sarah Ness
(Department of Civil and Environmental Engrg, University of South Florida, Tampa)
&
David R. Hokanson, PhD, PE, BCEE (Trussell Technologies, Inc., Pasadena, CA)

What also drives Water Reuse?

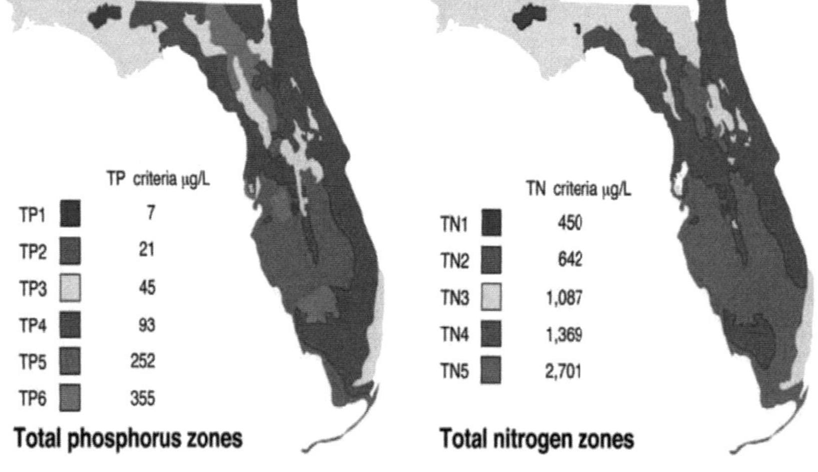

Learning Objectives

- Understand goal of our recent study for the WateReuse Foundation
- Explain difference between Embodied Energy, GHG Emissions, Carbon Footprint, and direct and indirect energy/emissions
- Understand what influences magnitude of CO_2 emissions and carbon footprints associated with water reuse
- Apply eGRID to calculate GHG emissions and carbon footprint for energy use from purchased electricity
- Be aware of available models to estimate GHG emissions and carbon footprint
- Differentiate between carbon footprint estimated using two models

Goal of Study we just completed for the WateReuse Foundation

- provide assistance to those who employ water reuse and desalination in estimating GHG emissions and carbon footprint
- recommend accessible models to utilities to provide estimations of GHG emissions and carbon footprint

Mihelcic, J.R., Zhang, Q., Hokanson, D.R., Cornejo, P.K., Santana, M.V., Rocha, A.M., Ness, S. J. (2013). "Feasibility Study on Model Development to Estimate and Minimize Greenhouse Gas Concentrations and Carbon Footprint of Water Reuse and Desalination Facilities," Project Report 10-12, 148 pages, WateReuse Research Foundation, Alexandria, VA.

Energy Consumption (e.g., direct energy) of Water Reuse and Desalination

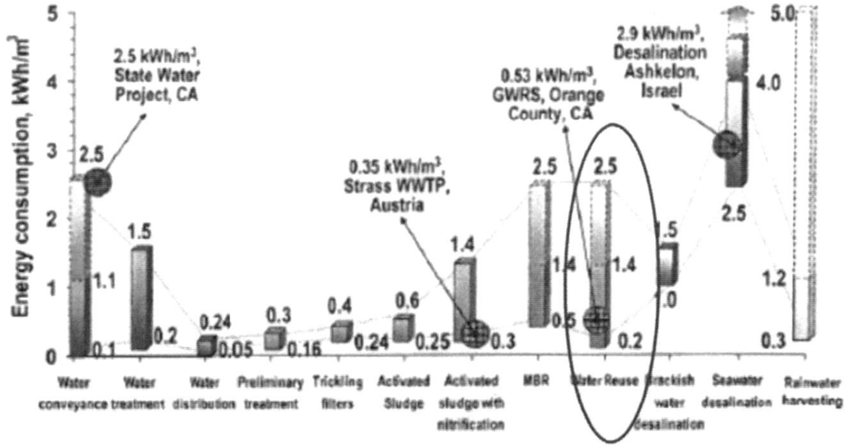

Source: Lazarova, Choo, and Cornel 2012

Lets Define Carbon Footprint, Direct and Indirect GHG Emissions and Embodied Energy

- A *carbon footprint* is defined as the total greenhouse gas emissions (reported in carbon equivalents) that are associated with a product, service, company, or other entity such as a household or water treatment plant. It consists of direct and indirect greenhouse gas emissions.
- *Direct emissions* are from sources owned or controlled by the reporting entity. *Indirect emissions* are a consequence of activities of the reporting entity, but they occur at other sources that are owned or controlled by another entity (Greenhouse Gas Protocol, 2012).

from Mihelcic, J.R., J.B. Zimmerman, *Environmental Engineering: Fundamentals, Sustainability, Design*, 2nd Edition, John Wiley & Sons, New York, 2013.

Direct & Indirect Energy when Reporting Greenhouse Gas Emissions and Carbon Footprint

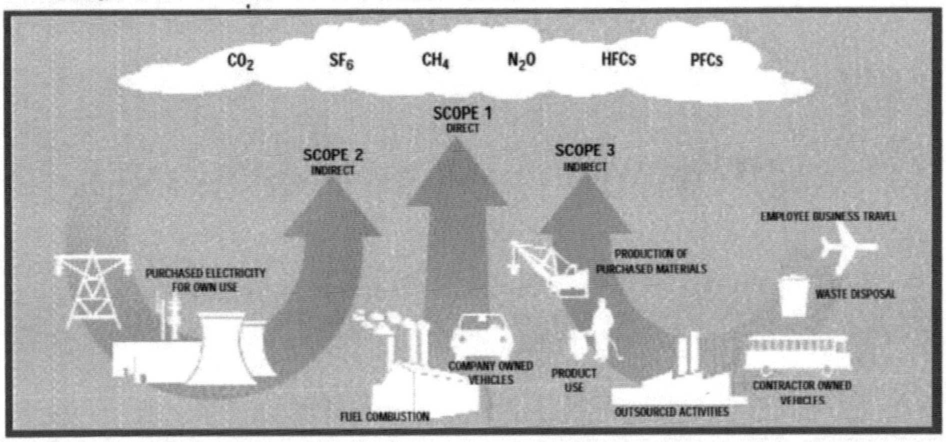

GHG Protocol Corporate Accounting and Reporting Standard

Life Cycle Assessment - Energy Impact

- LCA is a quantitative tool, which estimates the environmental impact of a system over its lifetime (EPA, 2006, Life Cycle Assessment: Principles and Practice)
- Embodied Energy – lifecycle energy consumption
- Carbon Footprint – lifecycle greenhouse gas emissions (GHGs)

Impact Category	Contributors		
Embodied Energy	N/A	Direct Energy (electricity)	Indirect energy: a) produce and transport materials to the facility, b) waste disposal, c) employee business travel.

CO_2 Emissions and Carbon Footprint of Water Reuse

Change in Emissions and Footprint with Capacity (per m3 of water treated)

Capacity (MGD)	CO_2 Emissions ($kg\ CO_2/m^3$)
0.8	0.2–1.1
13	0.4–0.8

Capacity (MGD)	Carbon Footprint ($kg\ CO_2\ eq/m^3$)
0.07, 0.62	0.1–0.9
1.3, 4.5, 5.6	0.5–1.2
10.6, 11.7, 26	0.1–2.4

Carbon footprint ranges from 0.1 to 2.4 kg CO_2eq/m^3

Carbon footprint per m^3 water produced appears to increase with increasing plant capacity (0.07 – 26 MGD)

Source: Mihelcic, J.R., Zhang, Q., Hokanson, D.R., Cornejo, P.K., Santana, M.V., Rocha, A.M., Ness, S. J. (2013). "Feasibility Study on Model Development to Estimate and Minimize Greenhouse Gas Concentrations and Carbon Footprint of Water Reuse and Desalination Facilities," Project Report 10-12, 148 pages, WateReuse Research Foundation. Alexandria. VA.

Change with Energy Mix - Facilities using renewable energy or energy mix with high portion of renewable energy have relatively low carbon footprint.

Energy Mix	CO_2 Emissions (kg CO_2/m^3)
Europe	0.8–1.0
France	0.23–0.27
New South Wales	0.4–0.8
Norway	0.14–0.16
Portugal	0.7–1.1

Energy Mix	Carbon Footprint (kg CO_2 eq/m^3)
Europe	1.3–1.9[1]
Israel	2.1
California	0.5–1.0
South Africa	0.1–0.7
Spain	-2.1–0.8
United States	1.7
Photovoltaic	0.2
Solar Thermal	0.1
Low Emissions[2]	0.9

[1]Based on Europe 2020 mix, which is composed of 35% renewable electricity production
[2]Low emissions refers to "a mix of renewable energy and current California sources" (Stokes and Horvath, 2009).

Source: Mihelcic, J.R., Zhang, Q., Hokanson, D.R., Cornejo, P.K., Santana, M.V., Rocha, A.M., Ness, S. J. (2013). "Feasibility Study on Model Development to Estimate and Minimize Greenhouse Gas Concentrations and Carbon Footprint of Water Reuse and Desalination Facilities," Project Report 10-12, 148 pages, WateReuse Research Foundation, Alexandria, VA.

Energy Mix: Emissions and Generation Resource Integrated Database (eGRID)

- eGRID provides conversion factors that allow a user to convert electricity usage (reported as MWh or GWh) to lbs of CO2, CH4, N2O, and CO2e.
- What is unique about eGRID is it makes this conversion using the energy mix that is unique to a particular region of the U.S. This is because the greenhouse gas emissions associated with electricity generation from consuming a particular amount of electricity differs around the country. This is based on a region's energy mix used to produce electricity that can consist of coal, natural gas, nuclear, hydro, biomass, wind, and solar.

Comparison of Greenhouse Gas Emission Rates in U.S. and several regions (data from eGRID2007 version 1.1, year 2005 data).

See http://www.epa.gov/egrid for data for all 26 U.S. subregions.

eGRID subregion name	CO_2 (lb/MWh)	CH_4 (lb/GWh)	N_2O (lb/GWh)	CO_2e (lb/MWh)
WECC California	724.12	30.24	8.08	727.26
SERC Virginia /Carolina	1,134.88	23.77	19.79	1,141.51
SERC Midwest	1830.51	21.15	30.50	1,840.41
FRCC all (Florida)	1,318.57	45.92	16.94	1,324.79
U.S.	1,329.35	27.27	20.60	1,336.31

Relating Individual Greenhouse Gas Emissions to Carbon Footprint

Greenhouse Gases (GHG)	GHG Emissions (kg)	Global Warming Potential*	Carbon Footprint (kg CO2 equivalent)
CO2	100	1	100
CH4	10	25	250
N2O	1	298	298

*IPCC Fourth Assessment Report 100-year time horizon

Total Carbon Footprint = 100 + 250 + 298 = 648 kg CO2 equivalent

eGRID does not account for line losses

- eGRID is based on generation of electricity and does not account for line losses from the point of generation to the point of consumption.
- Line losses range from 2.795% in Alaska, 3.691% in Hawaii, 5.333 in the Western U.S., 6.177% in Texas, and 6.409% in the Eastern U.S. (with a U.S. average of 6.179%).
- If a user wants to account for line losses in the estimation of greenhouse gas emissions, they would have to divide the eGRID generated greenhouse gas emissions by [1 – (percent line losses/100)] to determine the total greenhouse gas emissions that result from consumption of electricity

from Mihelcic, J.R., J.B. Zimmerman, Environmental Engineering: Fundamentals, Sustainability, Design, 2nd Edition, John Wiley & Sons, New York, 2013.

Example: Determine Carbon Footprint from Electricity Consumption Data

- Assume you own a building in Virginia or the Carolinas and you consume 11,000 kWh of electricity per year for heating, cooling, lighting, and operation of electronics and appliances. What is the amount of direct greenhouse gas emissions associated with CO2, CH4, and N2O (and the overall carbon footprint) for operating the building? Ignore line losses in your calculations.

from Mihelcic, J.R., J.B. Zimmerman, Environmental Engineering: Fundamentals, Sustainability, Design, 2nd Edition, John Wiley & Sons, New York, 2013.

Problem: 11,000 kWh of electricity per year. What is the amount of direct greenhouse gas emissions associated with CO2, CH4, and N2O (and the overall carbon footprint) for consuming this energy? Ignore line losses in your calculations.

eGRID subregion name	CO_2 (lb/MWh)	CH_4 (lb/GWh)	N_2O (lb/GWh)	CO_2e (lb/MWh)
WECC California	724.12	30.24	8.08	727.26
SERC Virginia/ Carolina	1,134.88	23.77	19.79	1,141.51
SERC Midwest	1830.51	21.15	30.50	1,840.41
FRCC all (Florida)	1,318.57	45.92	16.94	1,324.79
U.S.	1,329.35	27.27	20.60	1,336.31

Solution – Estimate GHG Emissions

- Using the conversion factors provided by eGRID (and listed in previous table for the sub-region of Virginia and the Carolinas), you can determine that the emissions of specific greenhouse gas emissions associated with operating this building as: 12,484 lb CO2, 261 lb CH4, and 218 lb N2O. There are 1,000 kW in 1 MW and 1,000,000 kW in 1 GW. These emissions do not account for line losses which are 6.409% in the Eastern U.S. To account for line losses, divide these eGRID generated emission values by (1-6.409/100).

from Mihelcic, J.R., J.B. Zimmerman, Environmental Engineering: Fundamentals, Sustainability, Design, 2nd Edition, John Wiley & Sons, New York, 2013.

- You can determine the carbon footprint by one of two methods. The easiest is to multiply the electricity consumption of 11,000 kWh by the CO2e conversion factor of 1,141.51 lb CO2e/MWh provided by eGRID (and listed in previous Table).

$$11{,}000 \text{ kW} \times 1{,}141.51 \text{ lb CO2e/MWh} \times \text{MW}/1{,}000 \text{ kW} =$$
$$= 12{,}556 \text{ lb CO2e}$$

This results in a value of 12,556 lb CO2e.

Problem: 11,000 kWh of electricity per year. What is the amount of direct greenhouse gas emissions associated with CO2, CH4, and N2O (and the overall carbon footprint) for consuming this energy? Ignore line losses in your calculations.

eGRID subregion name	CO_2 (lb/MWh)	CH_4 (lb/GWh)	N_2O (lb/GWh)	CO_2e (lb/MWh)
WECC California	724.12	30.24	8.08	727.26
SERC Virginia/ Carolina	1,134.88	23.77	19.79	1,141.51
SERC Midwest	1830.51	21.15	30.50	1,840.41
FRCC all (Florida)	1,318.57	45.92	16.94	1,324.79
U.S.	1,329.35	27.27	20.60	1,336.31

Solution – Estimate Carbon Footprint – Method 2

- You can find the solution in a longer manner, summing the contribution from each of the three greenhouse gases accounted for by eGRID, using the GWPs listed in Table.
- 11,000 kW x 1,134.88 lb CO2/MWh x MW/1,000 kW = 12,484 lb CO2 = 12,484 lb CO2e
- 11,000 kW x 23.77 lb CH4/GWh x GW/106 kW = 0.26 lb CH4 x 25 lb CO2e/lb CH4 = 6.5 lb CO2e
- 11,000 kW x 19.79 lb N2O/GWh x GW/106 kW = 0.22 lb CH4 x 298 lb CO2e/lb N2O = 65.5 lb CO2e
- The total GHG emissions in CO2e are the sum of these three values and equals 12,556 lb CO2e.
- Note the large amount of CO2 emissions from electricity generation here compared to the contribution of CH4 and N2O (even with their higher GWPs). This value is the carbon footprint of the building for one year when only considering direct emissions.
- Remember, these emissions do not account for line losses which are 6.409% in the Eastern U.S. To account for line losses, divide these eGRID generated emission values by [1 - (6.409/100)]. So footprint is now 13,416 lb CO2e

Previous example problem adapted from Rothschild et al., 2009, from Mihelcic, J.R., J.B. Zimmerman, *Environmental Engineering: Fundamentals, Sustainability, Design*, 2nd Edition, John Wiley & Sons, New York, 2013.

- You can now use e-GRID data and your specific electricity bills to determine GHG emissions and carbon footprint associated with your on site energy use.
- How about if you wish to consider emissions that consider the whole life cycle?

Available Models to Estimate GHG Emissions and Carbon Footprint

Practical Implications For Industry

- Limiting factor: data currently collected by industry
- Recommendation on data collection (at minimum)
  - information on electricity providers
  - the amount of water pumped and produced
  - facility-wide electricity usage
- Model development is needed
  - a user-friendly and robust model
  - applicable to different geographical regions
  - have an option that would require different levels of sophistication related to required input parameters

Method Used in Available Estimation Models

GHG Emission Estimation Method	Description of Methodology	Examples of Models that Fit this Methodology
Traditional LCA	Use process-based inventory	SimaPro, GaBi
Hybrid LCA-based models	Use both process-based and input-output-based inventory	Water Energy Sustainability Tool (WEST), WWEST, and WESTWeb
Specific models for estimating GHG emissions	Uses input parameters specific to user	Johnston Model, Tampa Bay Water Model
Other related models	NOT specifically used to estimate emissions from water reuse facilities, but contain aspects that are applicable	UKWIR Model, UK Environmental Agency Model, CHEApet, Systems Dynamics, GPS-X Model, mCO_2, Bridle and BSM2G

Hybrid Models use combination of EIO-LCA and Process-Based LCA

- **EIO-LCA** relies on national economic input-output (EIO) tables (showing relationship between different sectors of the economy) coupled with environmental impact tables to quantify metrics such as GHG emissions based on a set level of economic activity (i.e., the cost of equipment, pipes, chemicals)
- **Process-based LCA** is more detailed in that the environmental impacts are based on a specific analysis of the components in the system or product you choose to analyze. Emissions due to production of materials are calculated by EIO-LCA, emissions associated with energy production, transportation, and equipment usage are calculated using Process-Based LCA

Summary of Model Availability

Model Type	Emission Models	Tool Type	Available	Website or Contact Information
LCA-based models	SimaPro	Software	Commercially	www.pre.nl
	Gabi	Software	Commercially	www.gabi-software.com
	SiSOSTAQUA	Software	Commercially	www.simpple.com
Hybrid LCA-based	WEST	MS-Excel	Upon request	Dr. Jennifer Stokes at ucbwaterlca@gmail.com
	WWEST	MS-Excel	Upon request	Dr. Jennifer Stokes at ucbwaterlca@gmail.com
	WESTWeb	Web-based	Publically	west.berkeley.edu
Specific models	Tampa Bay Water	MS-Excel	Upon request	www.tampabaywater.org
	Johnston Model	MS-Excel	Upon request	Dr. Tanju Karanfil at tkaranf@clemson.edu
Other related models	CHEApet	Web-based	Publically	cheapet.werf.org
	UK Environment Agency Model	MS-Excel	Upon request	enquiries@environment-agency.gov.uk
	Bridle and BSM2G Models	Software	Publically	Author Lluis Corominas at lcorominas@icra.cat
	System Dynamics	Software	Commercially	www.iseesystems.com
	GPS-X	Software	Commercially	www.hydromantis.com/GPS-X.html
	Carbon Accounting Workbook, 5th version	MS-Excel	Commercially	www.ukwir.org
	mCO2	Software	Commercially	www.mwhglobal.com

Emission Sources Considered in Hybrid LCA and Specific Models

Emission Sources Considered	Hybrid LCA Models			Specific Models	
	WEST	WWEST	WESTWeb	Johnston Model	Tampa Bay Water Model
Material production	X	X	X		
Material delivery	X	X	X		
Electricity consumption	X	X	X	X	X
Electricity mix	X	X	X	X	X
Fuel use (on-site and fleet vehicles)	X	X	X	X	
Sludge disposal	X	X	X	X[1]	
Chemical production	X	X	X	X	
Direct process emissions		X[2]	X[2]	X[1]	
Process equipment			X[3]	X[4]	
Disinfection processes			X[3]	X[4]	

Applicability of Hybrid LCA and Specific Models

Applicability & Availability	Hybrid LCA Models			Specific Models	
	WEST	WWEST	WESTWeb	Johnston Model	Tampa Bay Water Model
Previously applied to water reuse	X				
Previously applied to desalination	X				X
Designed for wastewater facilities		X			
Designed for water facilities				X	
Designed for regional water supply					X
Currently regionally transferable	X	X	X	X	
Custom, state, and national electricity mix	X	X	X	X	
Available upon request	X	X	X	X	X

Comparison of WEST and Tampa Bay Models

Output Comparison of Carbon Footprint Using Tampa Bay Water and WEST Models

Facility	**Tampa Bay Water Model**	**WEST Model[1]**	**% Tampa Bay Water Model of WEST Model Estimate**
	CO_2eq (kg)/m³ Produced	CO_2eq (kg)/m³ Produced	
Desalinated seawater, membrane pretreatment	1.26	2.40	52%
Desalinated brackish groundwater	1.20	1.63	74%
Recycled water	0.53	1.02	52%

- Tampa Bay Water model: includes only electricity consumption
- WEST model: electricity consumption, fuel use by equipment and vehicles, chemical and material production

Example from: Mihelcic, J.R., Zhang, Q., Hokanson, D.R., Cornejo, P.K., Santana, M.V., Rocha, A.M., Ness, S. J. (2013). "Feasibility Study on Model Development to Estimate and Minimize Greenhouse Gas Concentrations and Carbon Footprint of Water Reuse and Desalination Facilities," Project Report 10-11, 148 pages, WateReuse Research Foundation, Alexandria, VA.

Tampa Bay Water Model

- Developed By Tampa Bay Water
  - Responsible for the extraction, treatment, and sale of water to member jurisdictions in the Tampa Bay metropolitan area
  - Model determines GHG emissions associated with water treatment of its facilities

Tampa Bay Water Model

Model Inputs	Units	Data Source
Water pumped	(MG/yr)	In-house
Water produced	(MG/yr)	In-house
Electricity Use from Pumping	kWh/yr	In-house
Gross load	MWh used/yr	U.S. EPA eGRID or CAM
CO_2 emission factors based on energy mix	Tons/yr	U.S. EPA eGRID or CAM
CH_4 emission factors based on energy mix	Tons/yr	U.S. EPA eGRID or CAM
N_2O emission factors based on energy mix	Tons/yr	U.S. EPA eGRID 2005
Electricity mix	% per source	U.S. EPA data and utility contacts

Model Outputs
CO_2 eq, CO_2, N_2O, & CH_4 emissions (lbs/kWh)
CO_2 eq, CO_2, N_2O, & CH_4 emissions (lbs/yr)
CO_2 eq, CO_2, N_2O, & CH_4 emissions (lbs/MG)

Tampa Bay Water Model

From internal operations data

1. Calculate Energy consumption per water produced

$$\frac{EnergyConsumed\ (kWh)}{WaterProduced\ (MG)} = \frac{EnergyConsumed\ (kWh)}{WaterPumped\ (MG)} * \frac{AnnualWaterPumped\ (MGY)}{AnnualWaterProduced\ (MGY)}$$

2. Calculate annual energy consumption

$$EnergyConsumption\left(\frac{kWh}{yr}\right) = \frac{EnergyConsumed(kWh)}{WaterProduced(MG)} * WaterProduced(MGD) * \frac{365d}{yr}$$

3. Convert annual energy use to emissions

From eGRID

$$Emission\left(\frac{lbs}{yr}\right) = EmissionFactor\left(\frac{lbs}{kWh}\right) * EnergyConsumption\left(\frac{kWh}{yr}\right)$$

Where $Emission\ Factor\left(\frac{lbs}{kWh}\right) = \frac{EmissionfromPowerProvider(\frac{tons}{yr})}{EnergyUsedbyPowerProvider\ (\frac{MWh}{yr})} * \left(\frac{2000lbs}{ton}\right) * \left(\frac{MWh}{1000kWh}\right)$

4. Convert amount of energy consumed/MG water to emissions/MG of water

$$Emission\left(\frac{lbs}{MG}\right) = EmissionFactor\left(\frac{lbs}{kWh}\right) * \frac{EnergyConsumed(kWh)}{WaterProduced(MG)}$$

Tampa Bay Water Model

Water Produced ↓ Energy Use ↓

	m3/yr	kWh/yr	MGD (Produced)	kWh/day	kwh/MG (Produced)	CO_2e lbs/day	CO_2e (lbs)/MG (Produced)	CO_2e kg/m3(Produced)
Water Supply								
Desal Seawater	*36000000*	*183,600,000*	26.0553	503013.6986	19305.6039	272681.7140	10465.4907	**1.256677932**
Desal Brackish	*36000000*	*103,680,000*	26.0553	284054.7945	10901.9881	153984.9679	5909.9241	**0.709653421**
Reused	*36000000*	*77,040,000*	26.0553	211068.4932	8100.7828	114419.3859	4391.4020	**0.527311917**

Adapted from Model provided by Tampa Bay Water

Inputs **Outputs**

Water Energy Sustainability Tool (WEST)

- Excel-based
  - Obtainable by contacting developers
  - http://west.berkeley.edu/model.php
- WESTWeb – Online interface
  - http://west.berkeley.edu/tool.php
- Water and Wastewater Treatment
- Hybrid-Life Cycle Assessment Based
- Calculates: CO_2-eq, NO_x, SO_x, PM_{10}, VOC, CO

WEST

- Composed of Environmental Assessment Methods
  - Economic Input Output (EIO)-LCA
    - Emissions from Materials Production
  - Process-Based LCA
    - Energy Generation
    - Transportation of Resources
    - Equipment Use

Framework of WEST Model

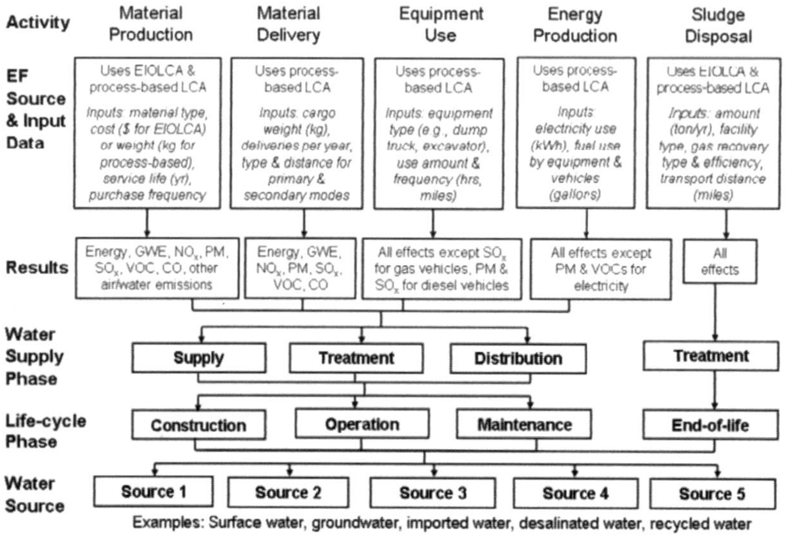

WEST

- The next few slides are to orient you with the WESTweb interface, as it is relatively easy to use.
- The website to access this program will be provided at end of talk. However, the Excel-based version, WEST, can be obtained by contacting the developers.

WESTWeb Interface

WEST Web

Modeling Parameters

Select system type: Water

Units selection: SI (metric)

Enter number of scenarios: 3

Functional Unit: 1000 (liters)

- Water or wastewater
- Units
- Number of Scenarios
- Scenario Capacities

Annual Water or Wastewater Production

Enter scenario production volume (in liters):

Scenario #	Scenario Name	Annual Scenario Production
1	Desalinated Seaw	36000000000
2	Desalinated Brack	36000000000
3	Recycled Water	36000000000

http://west.berkeley.edu/tool.php#results

WESTWeb Interface

Infrastructure

Pipe Length and Material

Would you like to enter detailed data about pipe materials? No

Enter total length (in meters):

Can include:

- Transport infrastructure information
- Material information (optional
- Processes (optional)

	Scenario #1 (meters)	Scenario #2 (meters)	Scenario #3 (meters)
Supply	3200	4800	1000
Treatment			
Distribution	1003000	1000000	35000

Reinforced Concrete Materials

Would you like to enter detailed data about buildings and pre-cast structures? No

Enter total volume of reinforced concrete (in cubic meters):

	Life (years)	Scenario #1 (cubic meters)	Scenario #2 (cubic meters)	Scenario #3 (cubic meters)
Supply				
Treatment	75			
Distribution				

Process Equipment

Would you like to enter detailed data about process equipment? No

WESTWeb Interface

Process Equipment

Would you like to enter detailed data about process equipment? Yes

Enter the dollars spent (in 2002$) on process equipment. Only include the purchase price; do not include labor or delivery costs.

Legend: ⇐ denotes supply, ⇄ denotes treatment, ⇒ denotes distribution.

Component	Life (years)	Scenario 1 (2002$) ⇐	Scenario 1 (2002$) ⇄	Scenario 1 (2002$) ⇒	Scenario 2 (2002$) ⇐	Scenario 2 (2002$) ⇄	Scenario 2 (2002$) ⇒
Filtration							
Filter Media (Sand or Gravel)							
Filter Media (Anthracite or Other Coal Product)							
Membranes							
General							
Pumps							
Fans / Blowers							
Motors and Generators							
Turbines							
Metal Tanks							
UV Lamps / Lights							
Other Industrial Equipment							
Electrical							
Controls							

WESTWeb Interface

Operation

Electricity Mix

Electricity Mix Location: US

Enter percentages for each scenario's electricity primary fuel/energy source:

	US Mix
Coal	49.61%
Oil	3.73%
Natural Gas	18.77%
Nuclear	19.28%
Hydro	6.5%
Biomass	1.3%
Wind	0.44%
Solar	0.01%
Geothermal	0.36%
Total	100%

State energy mix can be specified

http://west.berkeley.edu/tool.php#results

WESTWeb Interface

Energy Use

Enter quantities of energy consumed for each scenario:

*Use annual values

Annual Consumption of:	Scenario 1 ←	Scenario 1 ⇆	Scenario 1 →	Scenario 2 ←	Scenario 2 ⇆	Scenario 2 →	Scenario 3 ←	Scenario 3 ⇆	Scenario 3 →
Electricity (MWh)		13680	144000		9360	86400		16200	6840
Natural Gas (MMBTU)									
Gasoline (liters)									
Diesel (liters)									

Treatment Chemical Consumption

Enter quantities of chemicals used in each scenario:

	Units	Scenario #1	Scenario #2	Scenario #3
pH Adjustment				
Hydrochloric Acid	kg/yr	2916000	2340000	
Sulphuric Acid	kg/yr			
Lime	kg/yr			
Coagulants & Flocculants				
Aluminum Sulfate	kg/yr			1908000
Aluminum Hydroxide	kg/yr			
Caustic Soda	kg/yr		612000	
Ferric Chloride	kg/yr	14400		
Polymers	kg/yr			
Disinfectants				
Chlorine	kg/yr			684000
Calcium Hypochlorite	kg/yr	234000	396000	
Ozone	kg/yr			
Aqueous Ammonia	kg/yr	302400	468000	
Others				
Fluorosilicic Acid	kg/yr			
Other Chemicals	$/yr			

http://west.berkeley.edu/tool.php#results

WESTWeb Interface - Results

Scenario 3 Results
→ **Recycled Water**

				GHG g CO2e/FU	EN MJ/FU
INFRASTRUCTURE	Piping	Pipe	←	0	0
			⇆	-	-
			→	1	0
	Concrete and Buildings	Concrete and Buildings	←	-	-
			⇆	-	-
			→	-	-
OPERATION	Energy Use	Electricity	←	-	-
			⇆	296	4
			→	125	1
		Natural Gas	←	-	-
			⇆	-	-
			→	-	-
		Equipment Fuels	←	-	-
			⇆	-	-
			→	-	-
	Chemicals	pH Adjustment	⇆	-	-
		Flocculants / Coagulants	⇆	15	0
		Disinfectants	⇆	25	0
		Other	⇆	-	-
Σ	Infrastructure		←	0	0
			⇆	-	-
			→	1	0
	Operation		←	-	-
			⇆	337	5
			→	125	1
	End-of-Life		→	-	-
	Scenario 3 Grand Total			464	7

http://west.berkeley.edu/tool.php#results

Connecting Application Requirements and Fundamental Science: Delivering New Water Component Technologies

Steve Gluck[1], Guillem Gilabert Oriol[2], Katariina Majamaa[3], Santhosh Ramalingam[4], Abhishek Shrivastava[5]*

[1]Dow Water and Process Solutions R&D, *Freeport TX*; [2]Dow Water and Process Solutions R&D, *Tarragona, Spain;* [3]Dow Water and Process Solutions R&D, *Shanghai, China*; [4]Dow Chemical Core R&D, *Freeport TX*, [5]*Dow Chemical/Clean Filtration Technologies, Redwood City, CA*

Dow.com

Agenda

- Dow Introduction
- Dow Water & Process Solutions Overview
- Technology Screening
- Reverse Osmosis
- Ultrafiltration
- Macrofiltration

2

Dow Technologies and Market Solutions

The most complete portfolio in the industry today and a global presence second to none with #1 positions in Reverse Osmosis and Ion Exchange

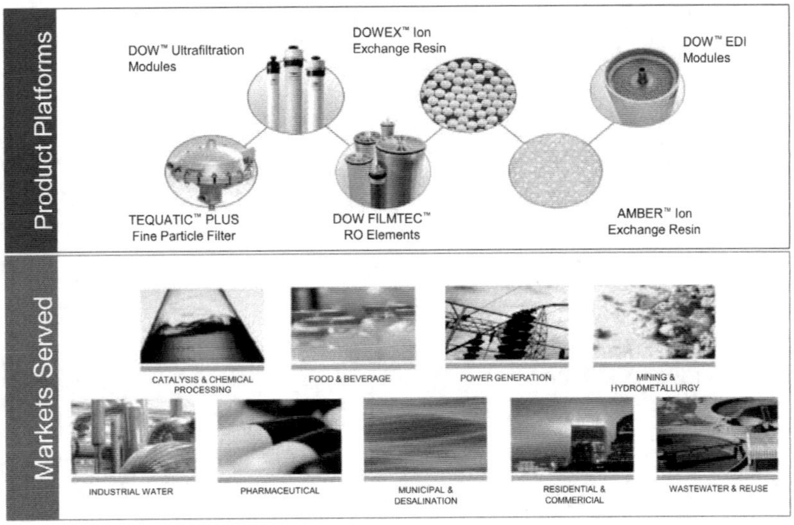

Enabling Components

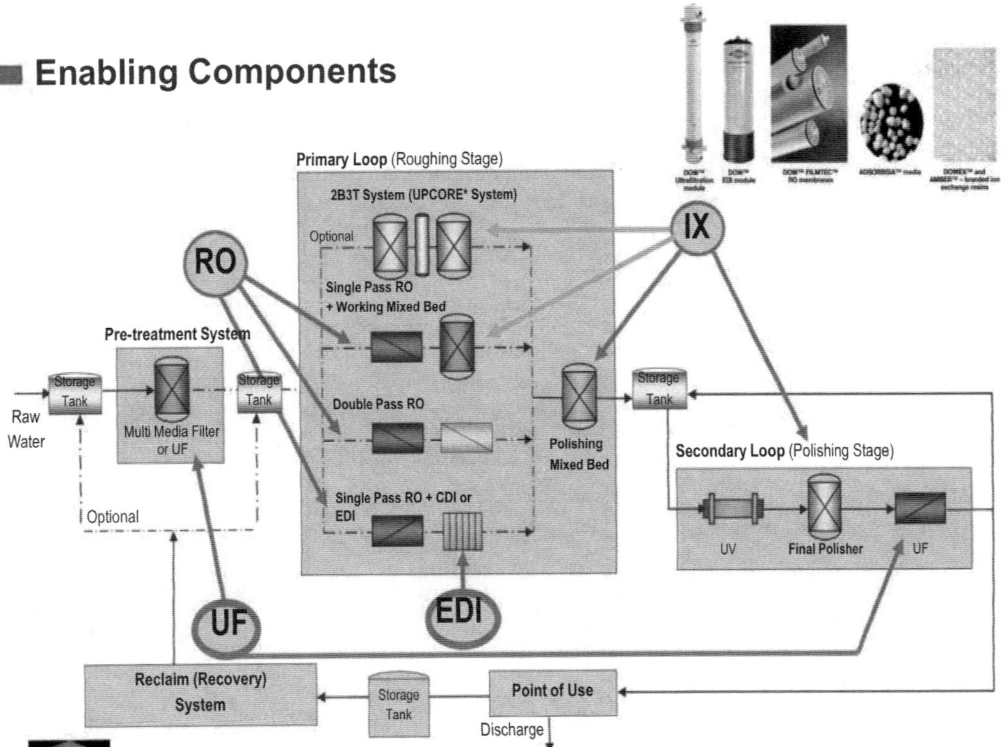

 4

Technology Screening

- Is it real?
  - Mass and Energy balance
  - Techno-Economic Analysis
  - Solid water/wastewater market drivers
- Does it fit?
  - Component, Equipment, System or Service model?
- Can it be more successful with a Dow relationship?
  - Augmented by R&D strength, global reach, partnership relations?
  - 1+1>2

RO and UF: Energy and Membranes

- Visible discussion because of OpEx considerations:
  - RO discussion has been largely around energy recovery devices and the membrane permeance
    - A high water permeability, "A" is desired
    - A low salt permeability, "B" is also desired
  - UF is concerned about operational flux
    - Fouling and subsequent water recovery issues

Specific Energy in Sea Water Desalination is Approaching the Practical Thermodynamic Minimum

Sea water desalination energy efficiency is asymptoting

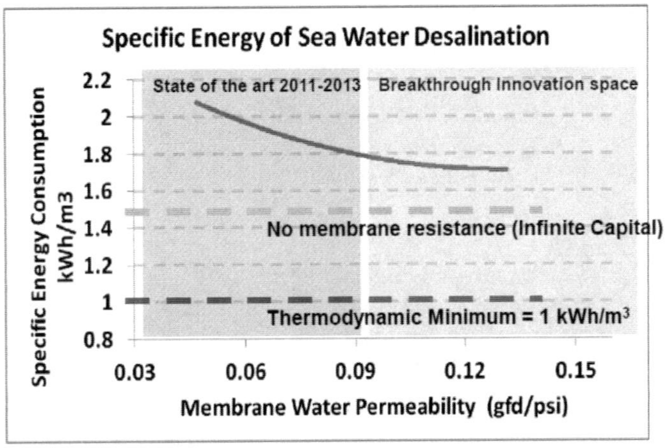

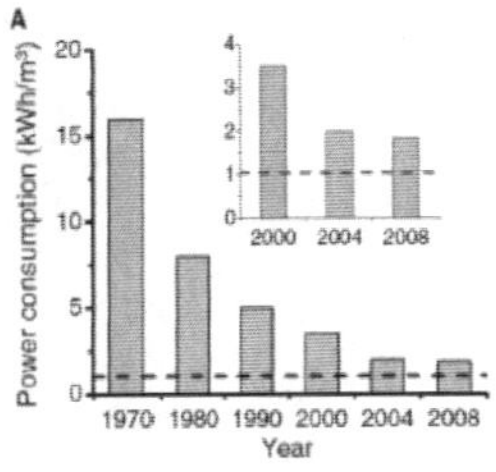

[1]Elimelech et al., Science 333, Aug 2011

DOW 7

Brackish Water Desalination Energy Breakdown

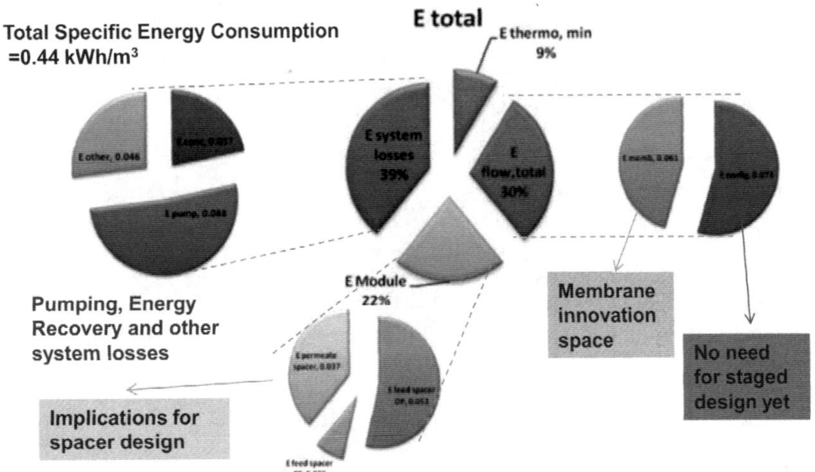

- Energy conversion losses ~ 40%
- Pressure losses in the module constitute 20% of the total energy consumption
- Membrane innovation potential before capital cost ~15% of total current specific energy

DOW 8

Specific Energy Trends in Brackish Water Desalination

Substantial improvements in membrane now are leading to smaller returns in energy savings due to other contributing factors

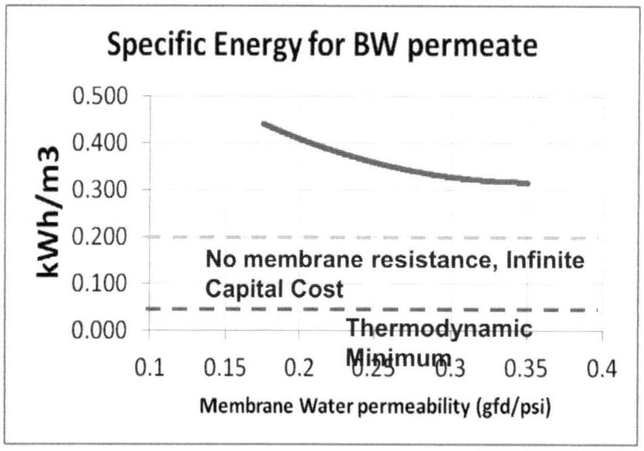

DOW 9

Current and Future Status

- Diminishing energy savings in desalination with membrane innovation alone
  - High permeability membranes alone will no longer give same returns in energy savings in reverse osmosis desalination as in the past
  - BW Desalination: System losses become a major portion of overall energy consumption
  - SW Desalination: Thermodynamic Minimum Energy for desalination accounts for more than half of the desalination energy
- Configurational energy (system design) and element design have elements that can enable further energy savings (at the cost of CAPEX)
- Research focuses on operational stability

DOW 10

Low Energy Brackish Water Membranes

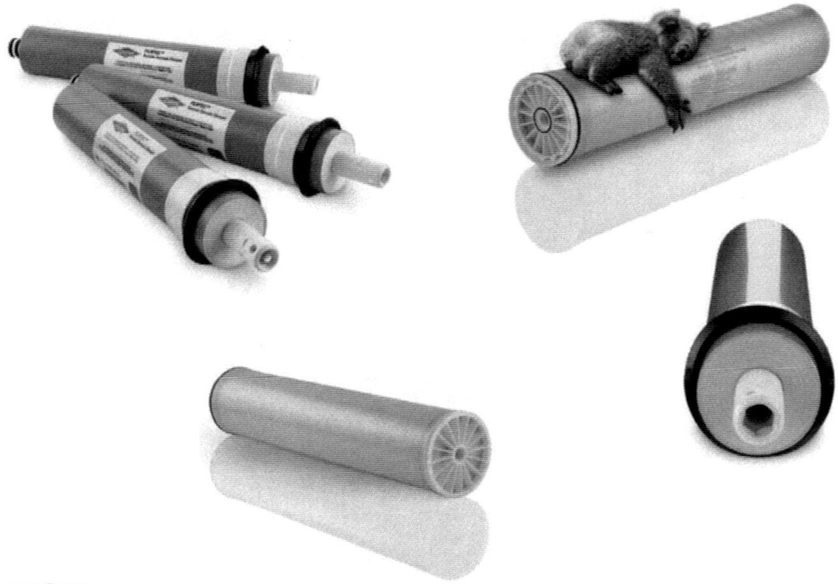

11

DW&PS Strategic Reduction Objective

Reduce cost of water by 35%

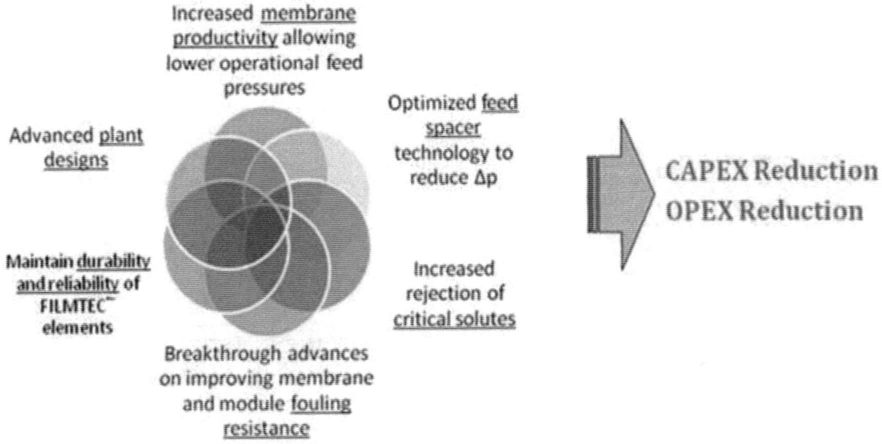

12

DOW™ FILMTEC™ Membranes – Novel Low Energy Membranes

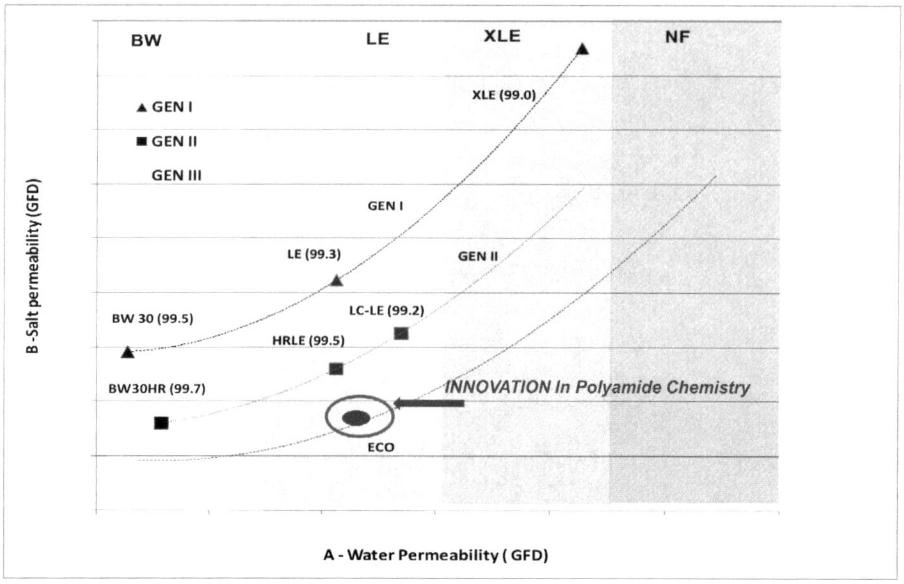

DOW

13

DOW™ FILMTEC Membranes – Novel Feed Spacer

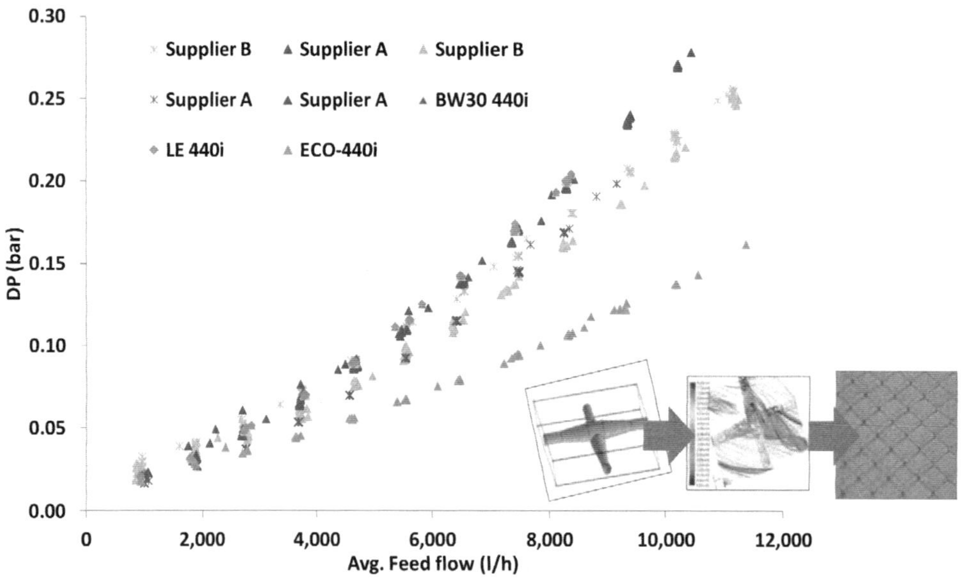

DOW

14

DOW™FILMTEC™ ECO Low Energy Brackish Water RO

- Delivers end-use customers the industry's highest rejection low energy BWRO membrane
  - Lower operating pressure by 30% while retaining permeate quality
  - Lower permeate TDS by 40% while maintaining feed pressure
- Target markets: Industrial & Power

Transformation With ECO, a new kind of plant emerges. One that runs on less energy and reduces environmental stress and strain. An element designed with an ecological conscience so powerful it transforms your plant into an entirely new ECO system. Lines blur between manufacturing plants and lush ecosystems. A phenomenon that causes naturalists to take note. And track ECO's effects accordingly.

Life Cycle: From Innovation to Product

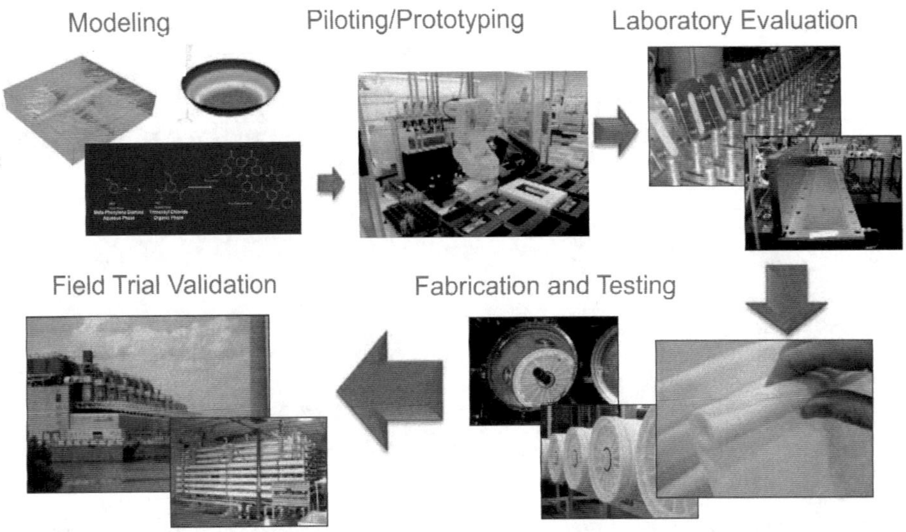

High-Performance DOW™ Ultrafiltration

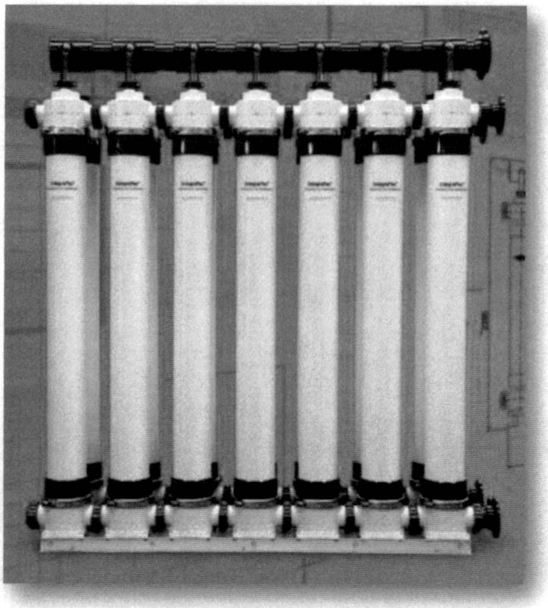

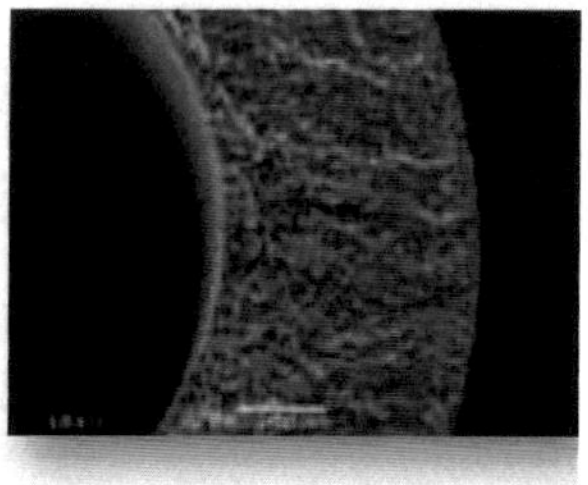

DOW IntegraPac™ Skids

17

Optimization of Ultrafiltration Membrane Cleaning Processes

Hypotheses:

The efficiency of the ultrafiltration process can be increased by improving its cleaning steps performance in SW Desal

Gilabert-Oriol, G.; Moosa, N.; Garcia-Valls, R.; Busch, M. and Garcia-Molina, V.; *Optimizing seawater operating protcols for pressurized ultrafiltration based on advanced cleaning research, Desal. and Water Treatment, 51(2013) 384-396.*

18

Fouling

- Fouling in ultrafiltration is due mainly by particles blocking membrane pores and particles building a gel layer above the membrane which reduces filtration area

- Therefore, to keep filtration flux constant, more pressure from feed pump (TMP) is required

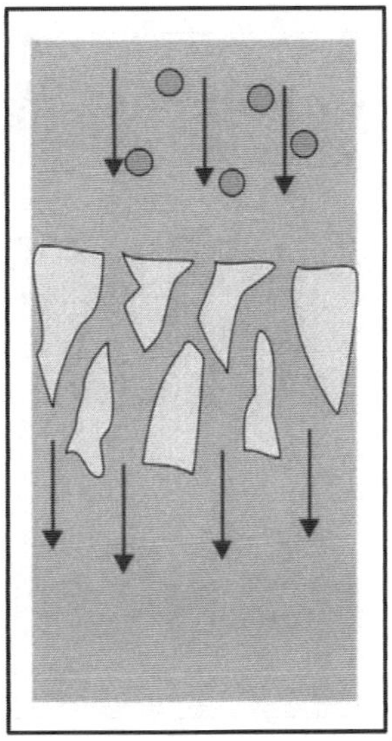

19

Cleanings

- Three different types of cleanings are done in order to maintain sustainable operation of ultrafiltration and thus controlling trans-membrane pressure

Cleaning	Foulant	Chemicals	Frequency
Backwash (BW)	Particle	-	Hours
Chemical Enhanced Backwash (CEB)	Biological	Chlorine	Days
Cleaning in Place (CIP)	Organic	Base	Month
Cleaning in Place (CIP)	Inorganic	Acid	Month

20

Backwash Contribution to Efficiency*

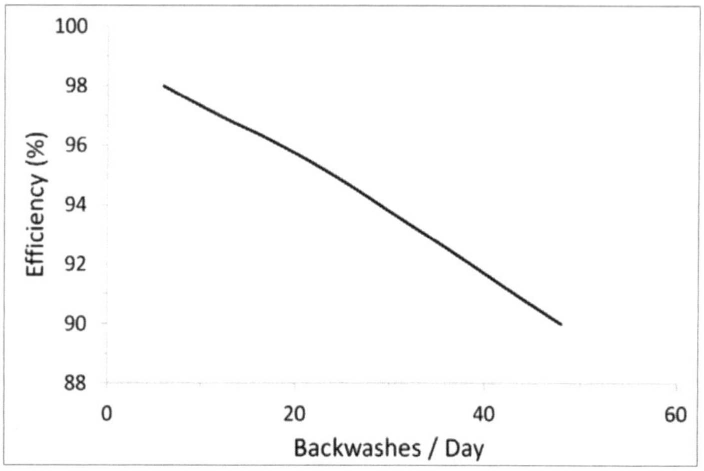

*Efficiency=Availability * Operational Recovery

21

Backwash Sequence

- Backwash sequence consists of five cleaning steps which are executed in order one after the other.

Order	Steps	Description
1	AS	Air Scour
2	D	Drainage
3	BWT+AS	Backwash Top with Air Scour
4	BWB	Backwash Bottom
5	FF	Forward Flush

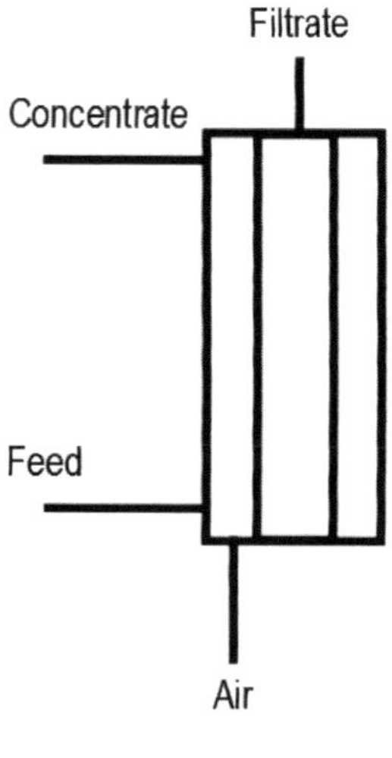

22

Air Scour

- Air Scour is used to shake the fibers creating a shearing effect between the different fibers which causes fouling to detach from membrane surface

Order	Steps	Description
1	**AS**	**Air Scour**
2	D	Drainage
3	BWT+AS	Backwash Top with Air Scour
4	BWB	Backwash Bottom
5	FF	Forward Flush

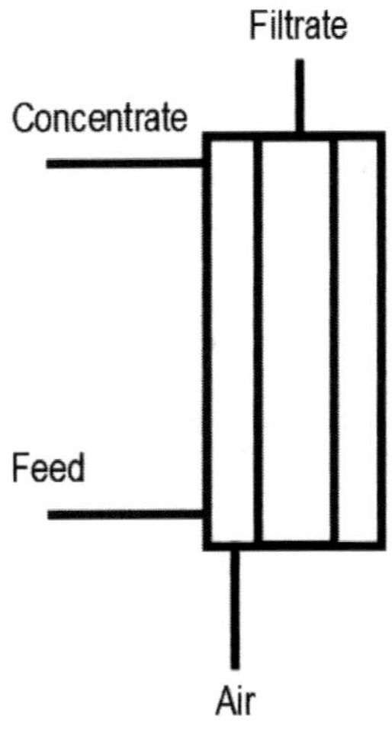

23

Drainage

- Drainage is used to empty the module after an Air Scour, so that all the dirtiness is removed.

Order	Steps	Description
1	AS	Air Scour
2	**D**	**Drainage**
3	BWT+AS	Backwash Top with Air Scour
4	BWB	Backwash Bottom
5	FF	Forward Flush

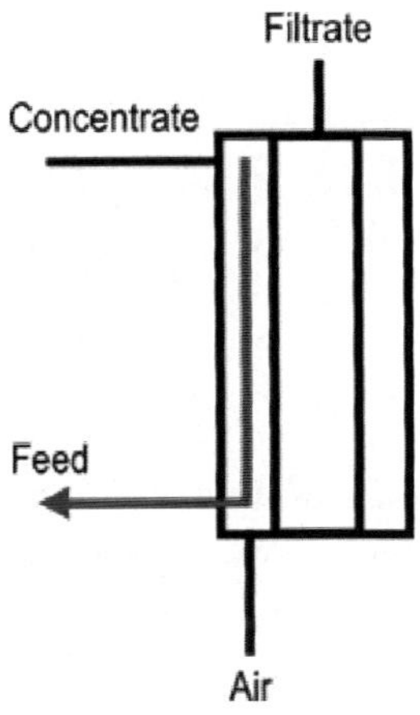

24

Backwash Top With Air Scour

- Backwash Top with Air Scour is used to force filtrated water to reversely pass through the fibers to unblock the pores and leave through the upper part of the module; air is fed to the module to create a shearing effect between the shaking fibers causing fouling to detach from membrane surface

Order	Steps	Description
1	AS	Air Scour
2	D	Drainage
3	**BWT+AS**	**Backwash Top with Air Scour**
4	BWB	Backwash Bottom
5	FF	Forward Flush

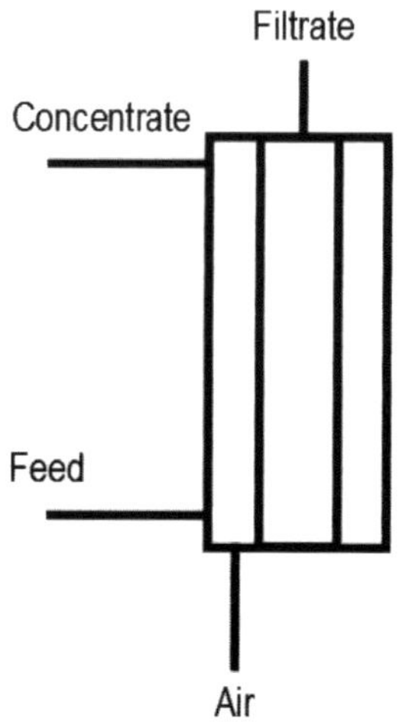

DOW

25

Backwash Bottom

- Backwash Bottom is used to force filtrated water to reversely pass through the fibers to unblock the pores and leave through the bottom part of the module

Order	Steps	Description
1	AS	Air Scour
2	D	Drainage
3	BWT+AS	Backwash Top with Air Scour
4	**BWB**	**Backwash Bottom**
5	FF	Forward Flush

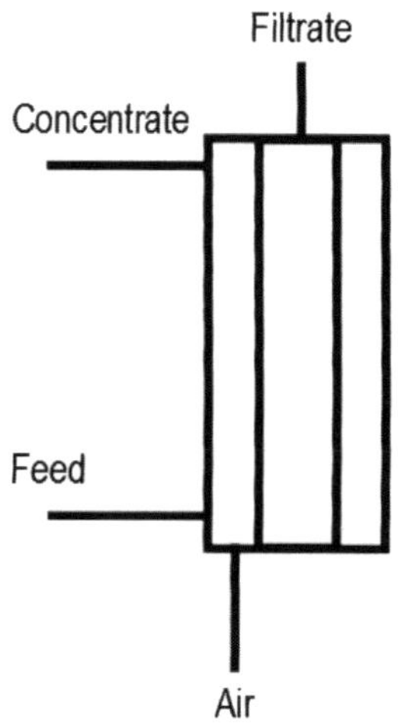

DOW

26

Forward Flush

- Forward Flush is used to create a shearing effect above the fibers so that fouling can detach the membrane surface

Order	Steps	Description
1	AS	Air Scour
2	D	Drainage
3	BWT+A S	Backwash Top with Air Scour
4	BWB	Backwash Bottom
5	**FF**	**Forward Flush**

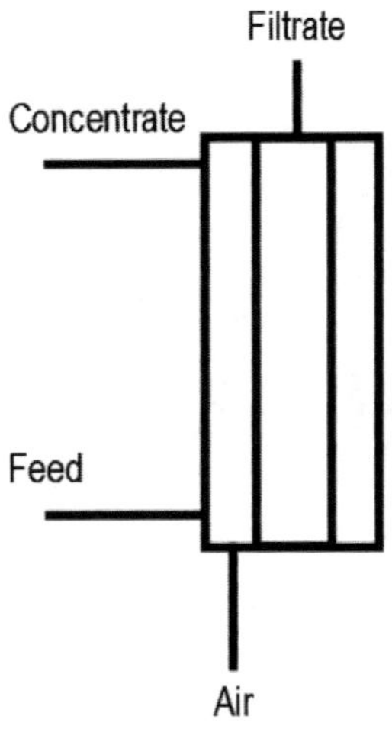

Desalination Plant

- Desalination plants integrate ultrafiltration pretreatment to remove particles such as colloids, algae, bacteria, viruses and macromolecules
- This allows operating reverse osmosis membranes at higher flux with less fouling. Therefore, reverse osmosis can focus on reducing salt content from water instead of also eliminating particles

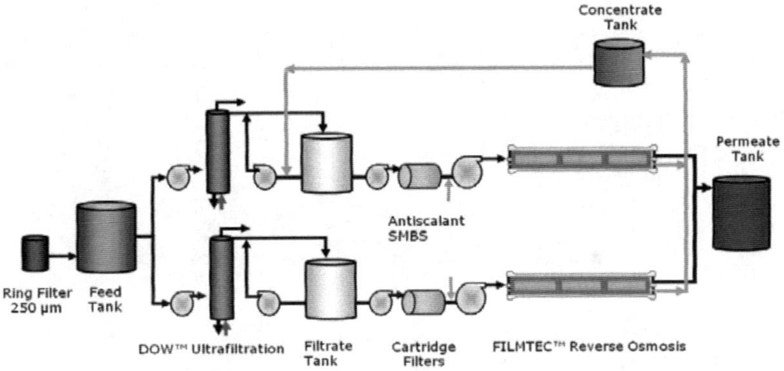

Results from Statistically Designed Experiment

- Backwash cleaning steps were reduced from 5 to 2
- Efficiency is increased from 88% to 95%, having the plant operating 96% of time with a 98% recovery
- Chemicals savings are increased to 9%

Exp	Frequency	Steps	Time	Recovery	Efficiency	Chemicals
-	30 min	5	91%	96%	88%	0.32 mg/l
1	30 min	2	96%	98%	95%	0.29 mg/l
Total	-	60%	5%	2%	7%	9%

DOW

29

TEQUATIC™ PLUS Fine Particle, Self-Cleaning Filters

- Nominal 10-50 um cutoff, 50-10,000 mg/L
- High recovery, low maintenance
- Outperforms competitive solutions for targeted applications
  - Produced Water disposal wells
  - Pre-RO treatment
  - Pulp & Paper, Food & Beverage

DOW

30

TEQUATIC™ PLUS Operation

1. Centrifugal force pushes dense particles outward
2. Vortex chamber continuously cleans filter
3. Solids are recirculated
4. Settled solids are purged

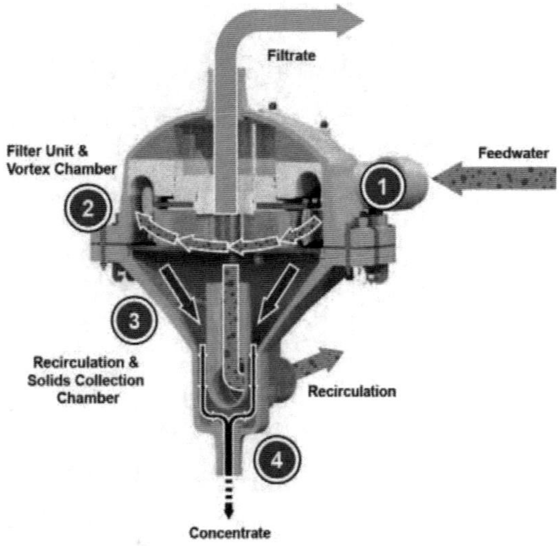

30 Micron Particle, 30 Micron Pore

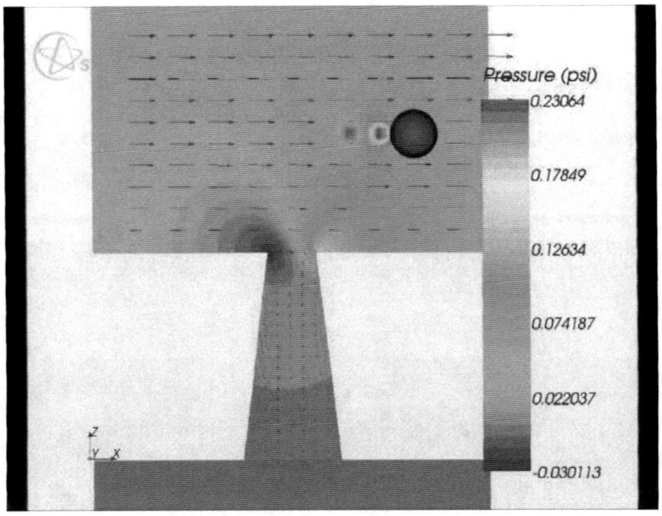

20 Micron Particle, 30 Micron Pore

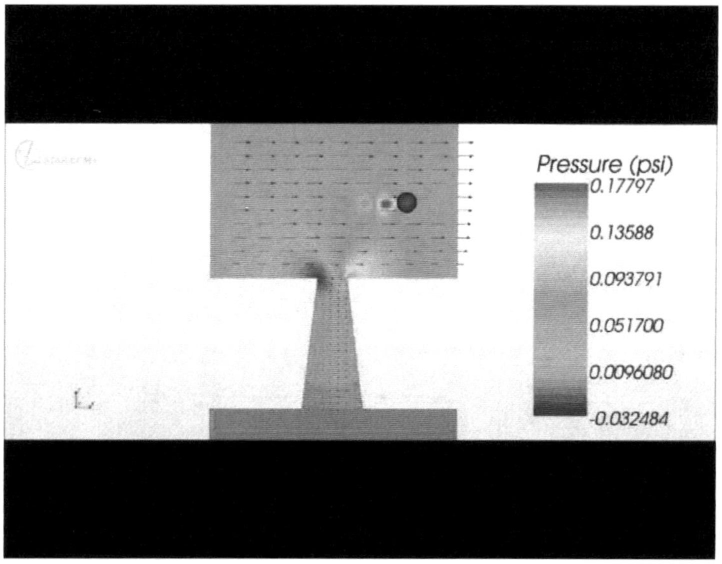

33

10 Micron Particle, 30 Micron Pore

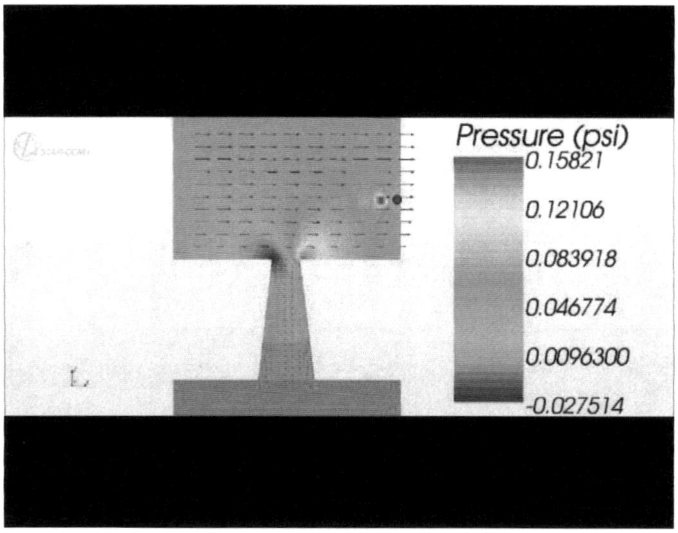

34

Summary, Examples and Expectations for the Future

- Refocused research on dominating forces within RO OpEx and CapEx
  - Fouling, configurations, element, selected rejection
  - Broader application requirements
- High operational flux UF membranes
  - Optimized cleaning
  - Higher solids
- Reliable, lower cost self-cleaning microscreens

35

Thank You

Dow Water & Process Solutions

Innovation that works harder to realize advancement in human progress

www.dowwaterandprocess.com

REUSE NOW

Industrial Water

Use and REUSE Workshop

ISWS San Antonio, TX
Larry Gurnari | Veolia Water Solutions & Technologies

VEOLIA
WATER
Solutions & Technologies

1

Presentation Overview

- OVERVIEW
- VEOLIA
- TECHNOLOGIES
- CASE STUDIES
- WATER IMPACT INDEX

REUSE NOW

- Something For Everybody:
  - Technologies
  - Unit processes/applications
  - Case studies including operating data
- Focusing on the different technologies and how they have been applied in REUSE applications world wide.
- No polynomials, organic chemistry (or inorganic chemistry for that matter)

REUSE NOW

- The US reclaims 6.5 % of treated (municipal) waste water ~(2.3 billion gals./day)
- Singapore reclaims 30%
- Israel reclaims 70%
- Australia reclaims 8%; but several projects on the board so this percentage will increase
- Industrial water usage growing at annual average rate of 3.7%
- Estimated annual industrial water withdrawals are in the 1.8 T gals (ie 10^9) (700 B m3)

REUSE NOW

- DEFINE REUSE?
  - Treat an "on site" source, use it on site and return water (discharge)
  - Treat an "off site" source (import) use it on site and return water
  - Treat on site and also treat discharge on site with no or some reduced return
  - Any treatment that reduces the amount of fresh water consumed while still meeting return limits (NPDES, POTW) for whatever might be discharged
  - Imperative no matter what conserve water at all stages
  - Crucial to work to together (Industry, Municipalities, Agriculture and Agentices)

REUSE NOW

- REUSE PROJECT SPECTRUM
  - Easy >>> Complex
  - Low Volumes (gpm) >>> MGD size
  - Less $ >>> $ MM
  - Weeks or months >>> Years

REUSE NOW

- IMPEMENTATION
  - Survey Plant (absolutely)
  - Bench Test (probably)
  - Pilot Demonstration Test (maybe)
  - Add/Change Chemistry
  - Add Equipment (refurbish)
  - Add Services (support, training, analytical)
  - Add Equipment, Chemistry & Services (bundled or wrap)
  - Change/Modify SOP's

Why REUSE

- Water Stress due to:
  - Population growth
  - Global climate changes
  - Industrial development
  - Droughts
- Cost & (un)reliability of supplies
- State regulatory will eventually enforce mandatory REUSE
- Pollution abatement; must treat anyway

Key Benefits Of Water REUSE

- Clients need to treat effluent anyway; why not recycle and reduce fresh water consumption
- Lowering fresh water consumption should lower operating water costs (water cost will increase over time as ***True Cost of Water*** is rationalized and accepted)
- Ability to expand production without increasing water consumption or negatively impact local enivronment
- Regulatory compliance and environmental sustainability working toward a ***Green Footprint***

The Paradigm Shift

- Waste Water >>> Water
- Recover Water
- Recover Energy
- Recover Resources

Scrap the Waste Water mentality

…(it's water); it may be off spec water; it may require treatment; but it is a recoverable and valuable resource

2

Veolia Environnement

- Organization
- Veolia Water
- Key Figures

Veolia Environnemnt

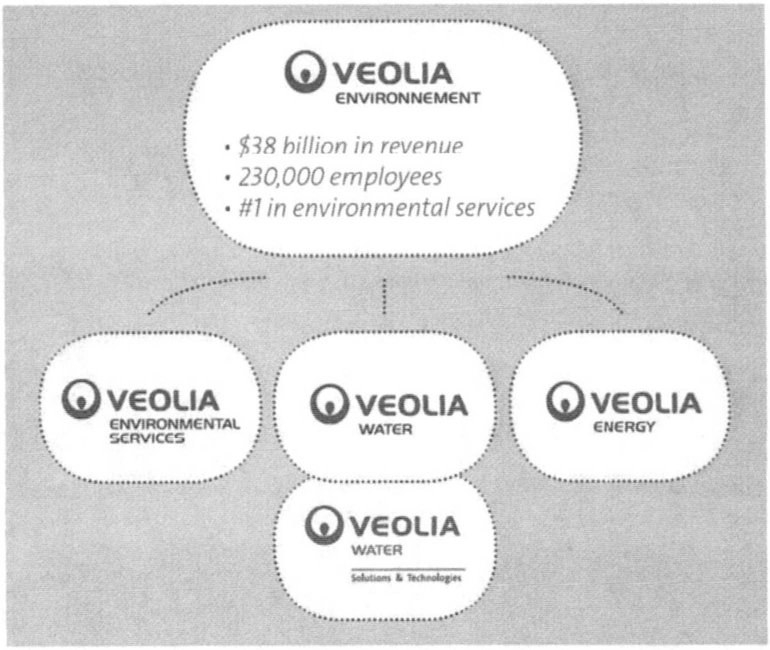

Veolia Water

- World Leader in Water and Waste Water
- Specialized in Outsourcing Services for
  - Municipal authorities
  - Industrial and service companies
- Major designer and integrator of technological solutions providing Water and Waste Water facilities and services worldwide

Key Figures

•89,094 Employees
•Provides water service to 100MM people
•Provides Waste Water service to 71MM people
•$15.9B revenue in 2012

Veolia Water Solutions & Technologies

- Global Network, Local Expertise
  - Technology and application development
  - Process design expertise; plant survey & evaluation; including bench-scale to full-scale pilot demonstrations
  - Pre-engineered and custom designed systems; variety of execution methods to suite the client including Mobile solutions
  - Hydrex™ chemistry/chemicals

Key Figures

•10,767 Employees
•Operating 54 countries
•Over 350 proprietary technologies
•$3.1B revenue in 2012

Veolia Water Solutions & Technologies

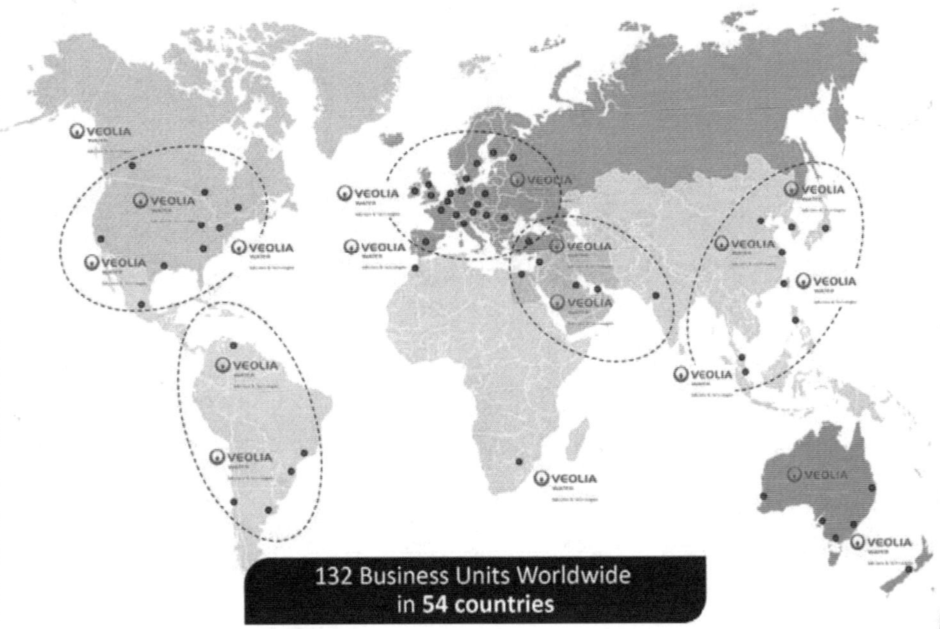

REUSE Design Considerations

- Sources
- Quality
- Users (Current & New)
- Drivers

REUSE Design Considerations

- SOURCES:
  - Surface, Groundwater, Well; (City)
  - Cooling Tower Blow Down
  - Boiler Blowdown
  - Stormwater/Rainwater
  - Waste Water (Municipal & Industrial REUSE are already inter-related and will become more so)

REUSE Design Considerations

- WATER QUALITY:
  - Total Dissolved Solids (TDS) cations/anion
  - Specific Cation/Anions (Ca + Mg, SO4, Cl, SiO2)
  - Organics (DOC/TOC; BTEX, PAH's, etc)
  - Trace Metals (As, Se, Fe, Mn, Al, etc)
  - Possibly Biological (if so BOD, COD, nitrogen, pathogens, nutrients, etc)

REUSE Design Considerations

- CURRENT RETURN POINTS:
  - NPDES
  - POTW
  - Impoundment (keep on site)
  - Deepwell Disposal
  - ZLD (maybe no return of any liquid)
- NEW USERS:
  - Rates, Volumes
  - Water Quality
  - Water Stress
  - Other Options (conservation, etc)

Drivers for REUSE

- Source, Quality, Return and REUSE options drive Technology development and application knowledge
- Existing technologies, expertise and experience do and will meet water quality challenges
- The value of water will increase significantly in the future (***"True Cost of Water"***)
- Local infrastructure requirements could limit reuse opportunities (location issue)
- Many locations (expecially coastal) experiencing severe water shortages (poulation growth, climatic)
- Many existing water management strategies are unsustainable

4

Technologies

- Summary
- Main Design Features

Key Technologies

- ACTIFLO® (TSS, turbidity, softening, TOC reduction, etc.)
- MULTIFLO™ (higher TSS applications)
- Discfilter
- AnoxKaldnes™ Moving Bed Biofilm Reactor (MBBR)
- NEOSEP™ Immersed Membrane Bioreactor (MBR)
- OPUS™ & OPUS II Technologies (water recovery)
- Zero Desalination Discharge (ZDD) (water recovery; brine management as well)
- DenseSludge™ (HDS) (reduction of liquid sludge)
- New Patent on Low Level Phosphorous Removal

Available Technologies

- *Suspended Solids Removal*
  - Actiflo®
  - Multiflo®
  - DenseSludge® and Multimedia Filters
  - 3FM Filters® (Flexible Fibre Filter Module)
  - Precoat Filters
  - Membranes (MF, NF, Ceramic)

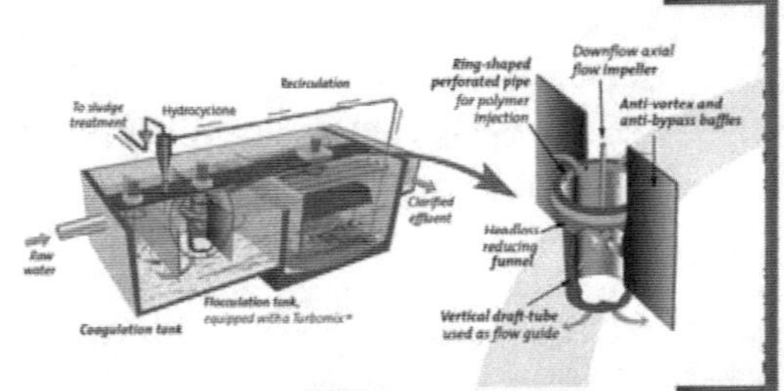

- *Dissolved Organics and Inorganics Removal*
  - Macro Porous Polymer Extraction (MPPE®)
  - Granular Activated Carbon (GAC)
  - Reverse Osmosis (RO)
  - Nano Filtration (NF)
  - Membrane Degasification
  - Desalination

- *Oil Removal*
  - CPI, DAF, IGF, TiPSS®
  - Walnut Shell Filter : Power Clean®
  - MPPE® (L/L extraction; BTEX, PAH)
  - Ceramic Membranes

Available Technologies

- *Demineralization*
  - Reverse Osmosis
  - Ion Exchange
  - Condensate polishers
  - Mixed bed polisher
  - Chemical dosing systems
  - Deaerators

- *Biological Treatment*
  - AnoxKaldnes® MBBR
  - MBR: Biosep®
  - Fluidized Media Fixed film
  - Off-gas BioFilter
  - VOC control

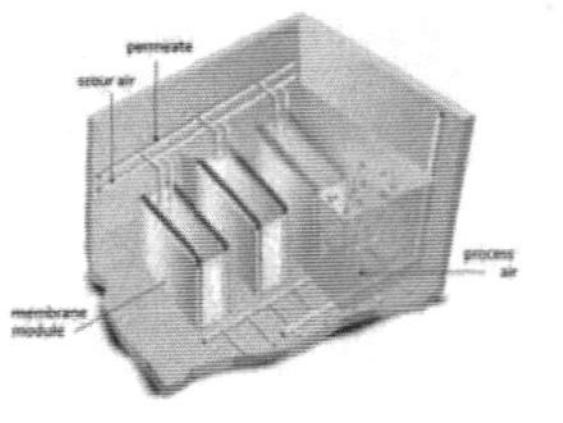

Available Technologies

- ***Salt Removal***
  - KOH Recovery (Hardtac®)
  - Multi Stage Flash (MSF) evaporation
  - Multi Effect Distillation (MED)
  - Reverse Osmosis (RO)
  - Mechanical Vapour Compression (MVC) Evaporation
  - Brine Concentration / Crystallization (ZLW or ZLD)

- ***Cooling Water Treatment***
  - Side-stream filtration
  - Cooling tower blow down treatment
  - Zero liquid discharge and reuse

- ***Mobile Water***
  - Aquamove® / Aquadem®

- ***Specialty Water Treatment Chemicals***
  - Hydrex® products

Available Technologies

- ***Sludge Treatment***
  - Thickening
  - Filter press and belt press dewatering
  - 3 and 2 phase centrifuges
  - Drying
  - Incineration (Pyrofluid®)
  - Coker injection
  - Odor detection and control: Odowatch®

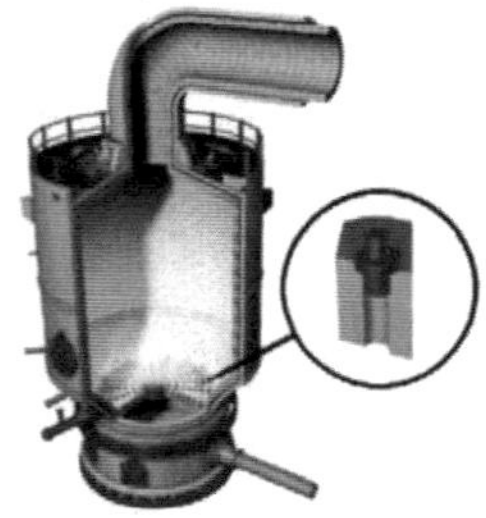

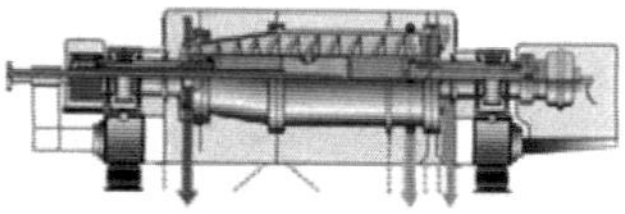

- ***Demineralization (Ion Exchange)***
  - Single and Mixed Beds
  - WAC/SAC – WBA/SBA
  - Selective Ion Exchange

ACTIFLO® Ballasted Flocculation

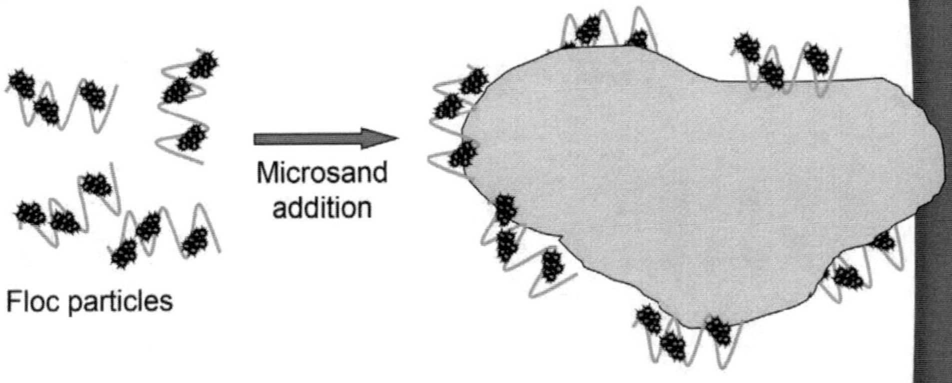

ACTIFLO® Principle

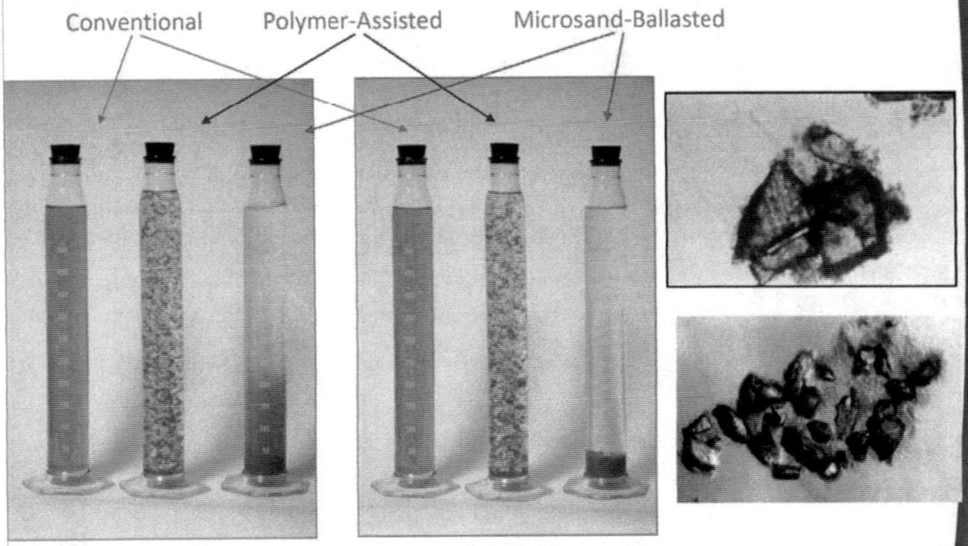

3 seconds **8 seconds**

ACTIFLO® Flow Schematic

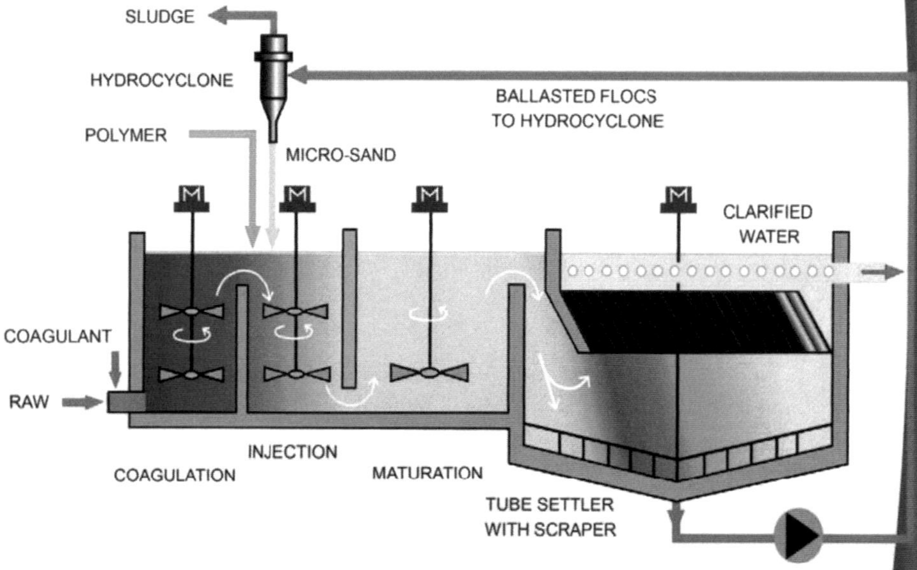

ACTIFLO® Process
Higher Rates Result in Lower Footprints

Process >>>	Conven-tional	Lamella Clarif.	Ballasted Floc.
Rise Rate (gpm/ft²) >>>	(0.2 - 0.8)	2.4 - 6.0	16.0 - 48.0
Footprint >>>		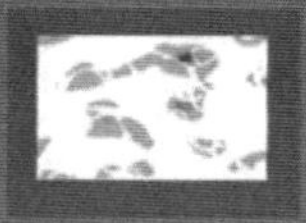	

Actiflo ThyssenKrupp Calvert, AL

MBBR (Moving Bed Bio-Reactor)

Characteristics

- AnoxKaldnes™ is the MBBR pioneer
- Biological treatment of carbon and/or nitrate using biomass attached to a moving carrier
- Continuous process, no backwash
- Compact, small footprint

Advantages

- High tolerance to load and toxicity variations
- Lower solids production (loading on clarifier)
- Better removal efficiencies at higher loading rates
- Tolerates lower operating temperature than conventional suspended growth systems

MBBR - Applications

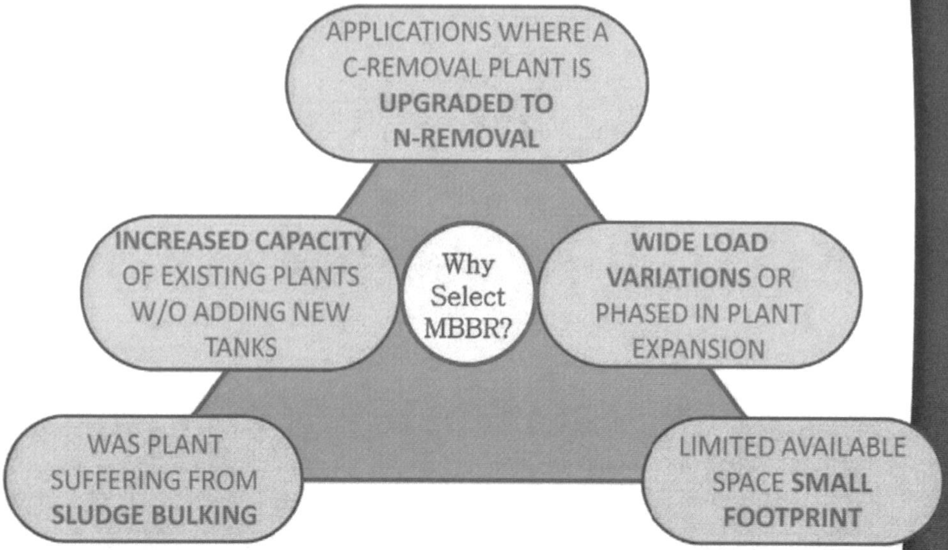

▶ **INSTALLED BASE**
MORE THAN 500 UNITS IN OPERATION

MBBR – Footprint Comparisons

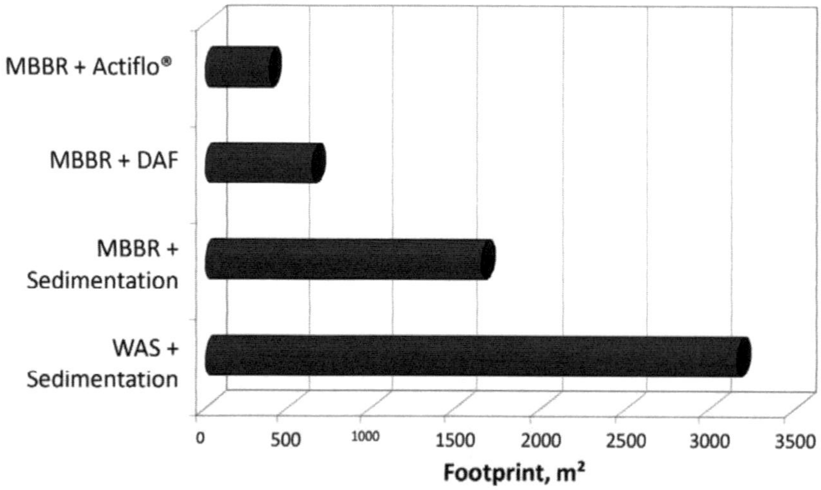

MBBR Process Configurations

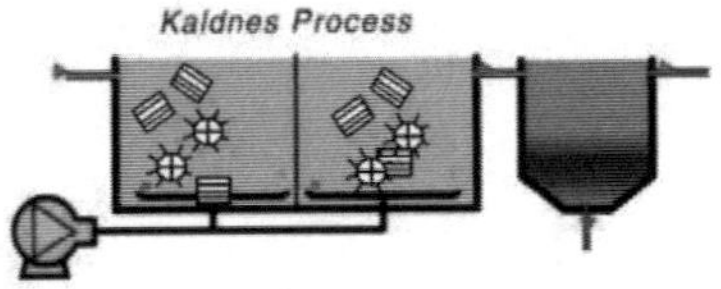

MBBR Process
(standalone no RAS; can be multiple reactors depending on BOD, Nitrify, De-nitrify, TSS loading)

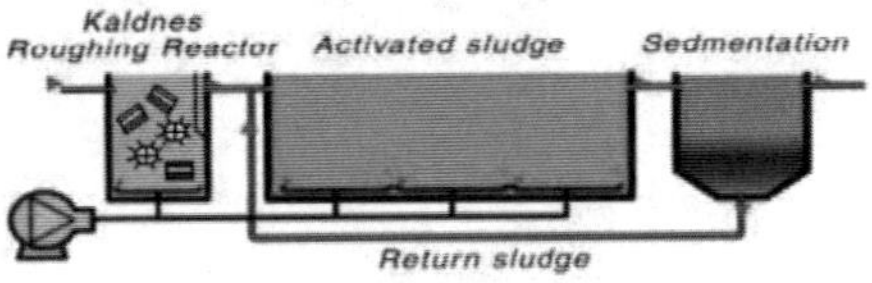

BAS™ Process
(Improve perf. or increase capacity of existing plant)

HYBAS™
(BOD >>> BOD + N)

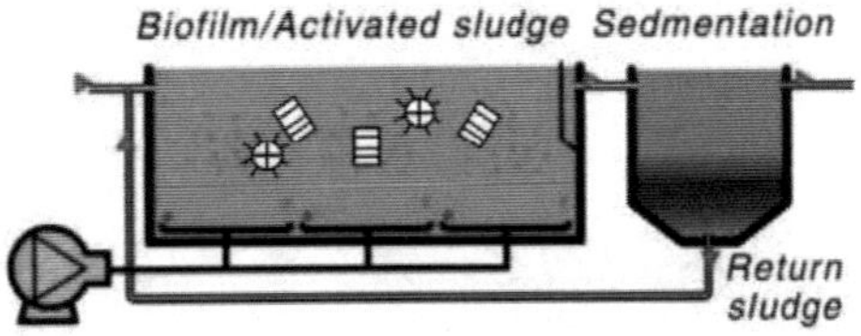

Discfilter

- Unique Filtration Technology
  - Woven Cloth Filter Elements
  - Inside -Out Pattern; Gravity Flow
  - Counter-Current Backwash Process
  - Particulate Removal
- High Rate Filtration Process
  - 70 to 80% smaller than Conv. Process
  - Small Footprint Modular Design
  - Low Capital & Installation Costs
  - Low Backwash Waste Volumes
  - Low Head Loss 8" – 12"
- Options
  - Stand alone Package Technology
  - Placement in Concrete Tanks
- Good Quality Effluent
  - Turbidity < 2 NTU

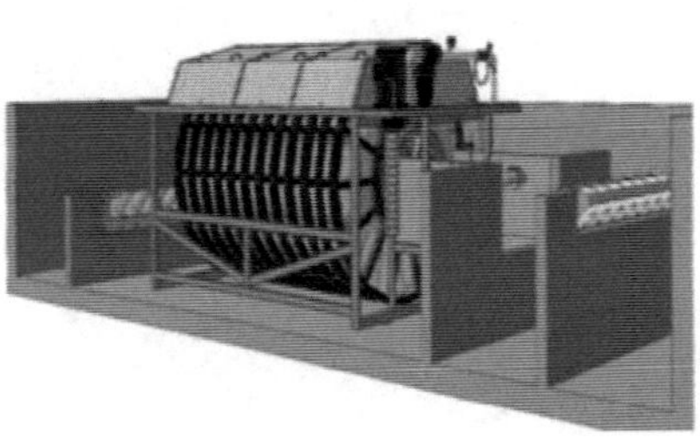

Discfilter

- Clarifier Polishing
- CT Make-Up & CT Sidestream Filtration
- Effluent Polishing; Title 22 Certified for Tertiary Filtration
- Gravity flow; rates up 6000 gpm per unit; multiple trains
- Integral Backwash Pump
- 10 -50 micron polyester replaceable woven screens

CeraMem Ceramic Membrane Module

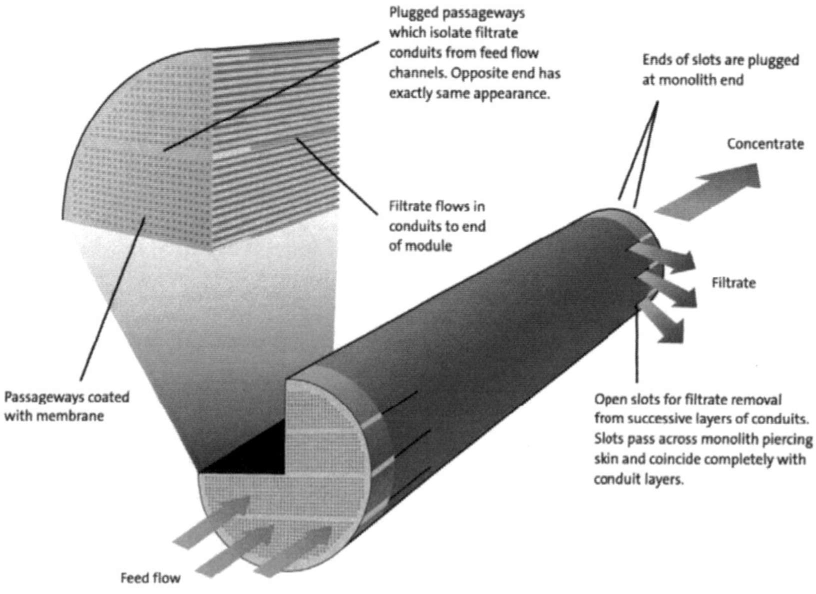

CeraMem Ceramic Module Operation

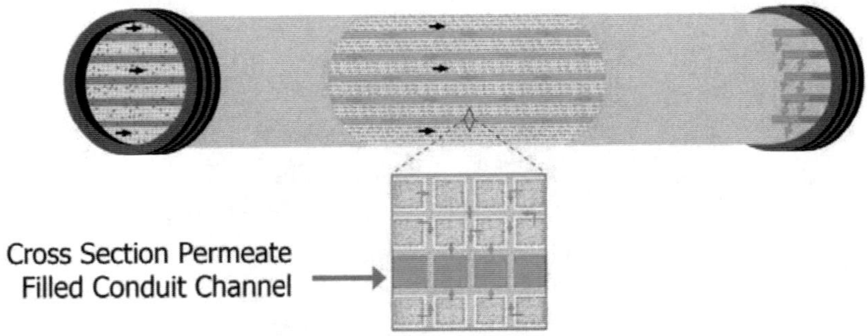

CeraMem Ceramic Module Cross Section

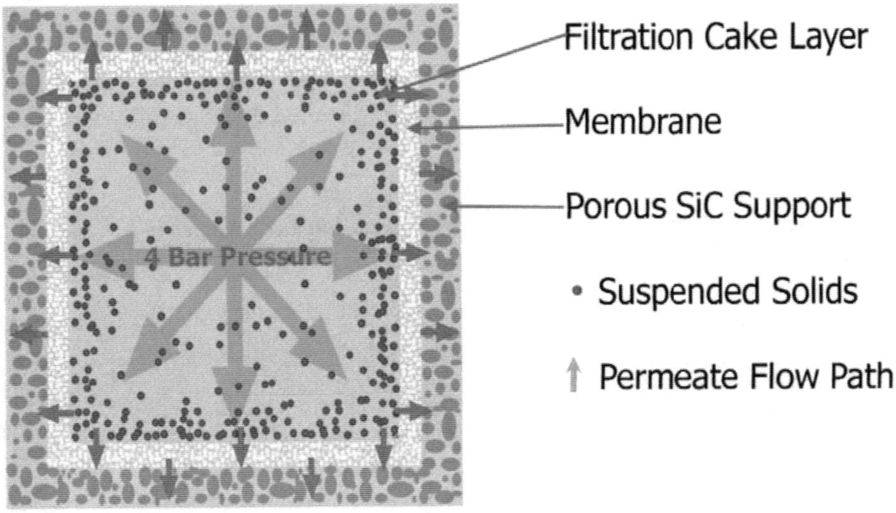

SAGD OPUS Pilot Results
Permeate, Feed, and Concentrates Samples

Brine Management

- Evaporation Ponds
  - Large Land Area; Capital Intensive (construction, liners, permitting)
  - Limited to Arid Climates Only
- Deep Well Injection (DWI)
  - Often Difficult To Permit
  - High Pumping and Monitoring Costs
  - Possible Contamination of Local Water Supplies; Deteriorate Soil Quality
  - Can be costly; level of risk to users
- Hauling
  - Expensive (Up to $0.25/gallon)
- Sewer Disposal
  - Salt impacts downstream treatment facilities
- Zero Liquid Discharge (Evaporator/Crystallizer)
  - Capital and Operation Cost Intensive for large volumes

Optimized Pretreatment & Unique Separation OPUS™

- TDS, Metals & TSS Reduction
- Tower Blowdown
- ZLD Pretreatment
- 90%+ Recoveries

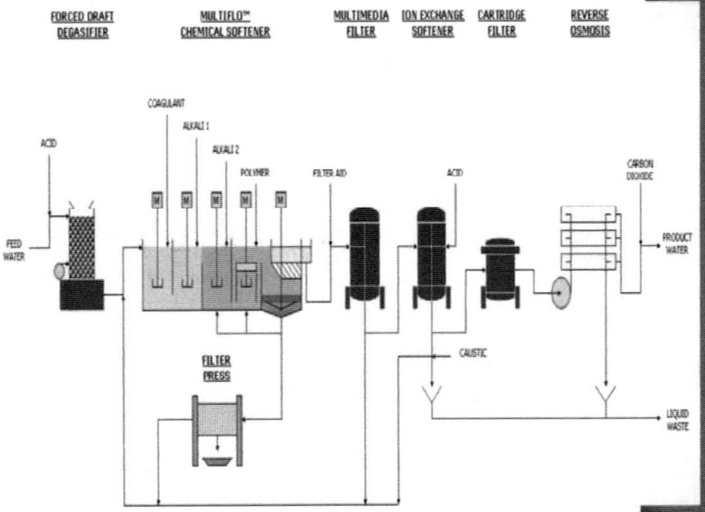

1st Generation OPUS

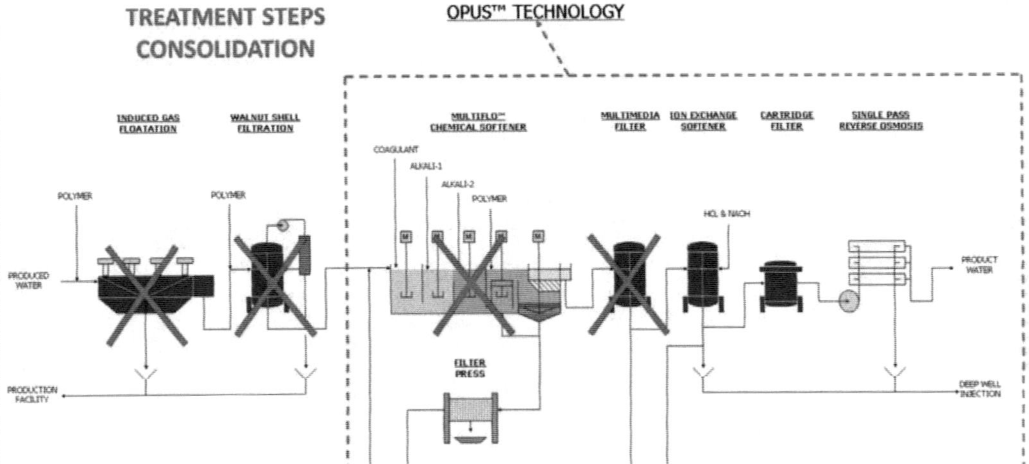

OPUS® II Technology

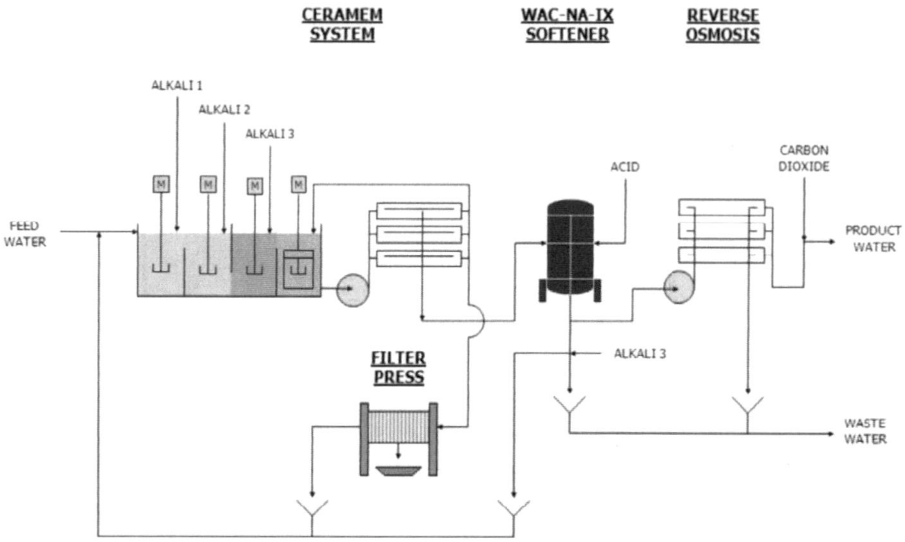

45

Zero Desalination Technology (ZDD)

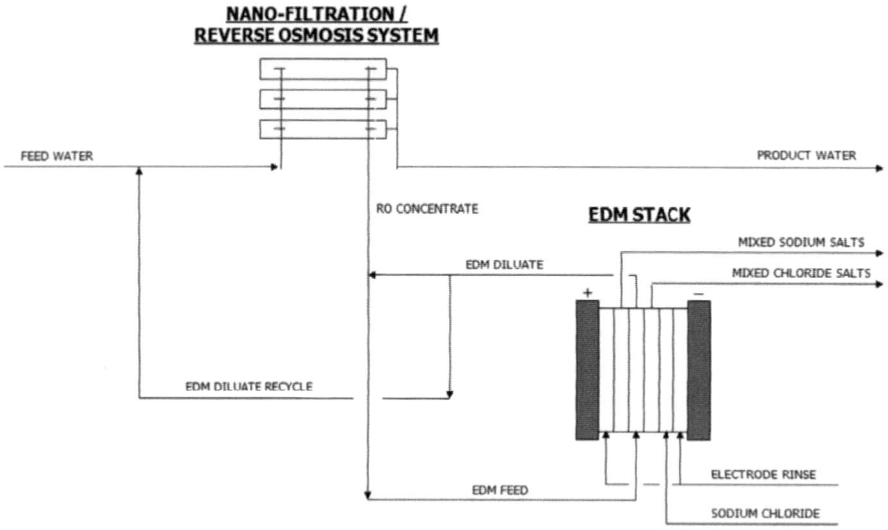

Reverse Osmosis - Benefits

- Reliable and Consistent Technology
- Easy Operation & Maintenance
- Treated Water Meets Potable Quality Standards
- Low Chemical Consumption
  - High Salt Rejection (> 98%)
  - Effective at Removing Emerging Compounds
  - Membrane is a Physical Barrier

Reverse Osmosis - Considerations

- Brine Management (especially as the feed TDS increases)
  - <100% Conversion (75% Typical)
  - Salinity
  - Sustainability
- Energy if Costs Rise
- Scaling Limits Efficiency and Recovery (if it cannot be controlled)
- Pretreatment can be >50% of cost ($SDI_{15} < 3$)
- Environmental Issues for Concentrate Return
- Challenging Treatment Chemistry Especially for Potential REUSE recycle applications
  - Silica, Sulfate, Carbonate, etc.
  - Impacts Extent and Cost of Recovery

Need for Economical High-Recovery Water Systems

The ZDD Process & How it Compares to RO?

- RO or NF
  - Pressure driven process moving ions across a semi-permeable barrier to reduce TDS
- ED
  - electricity is the driver to move ions across a membrane barrier again reducing TDS

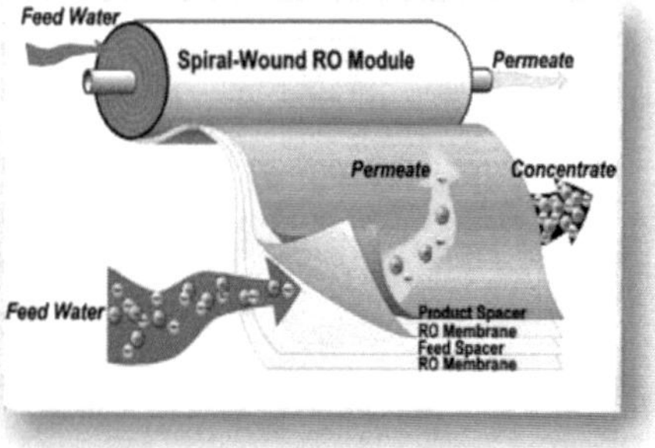

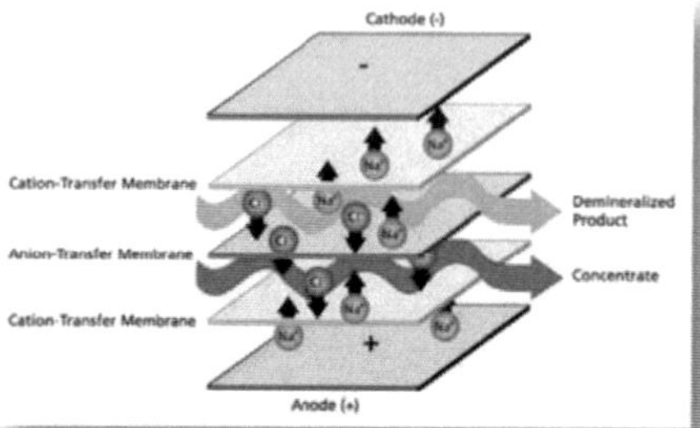

Electrodialysis Metathesis (EDM)

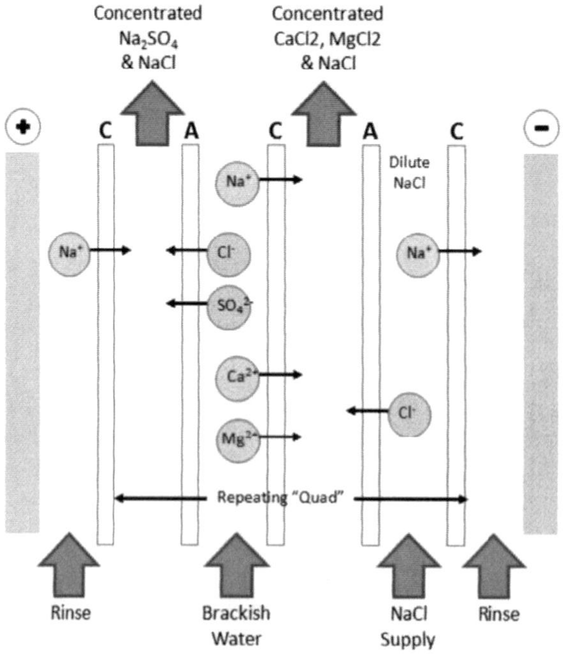

ZDD Process Flow Diagram

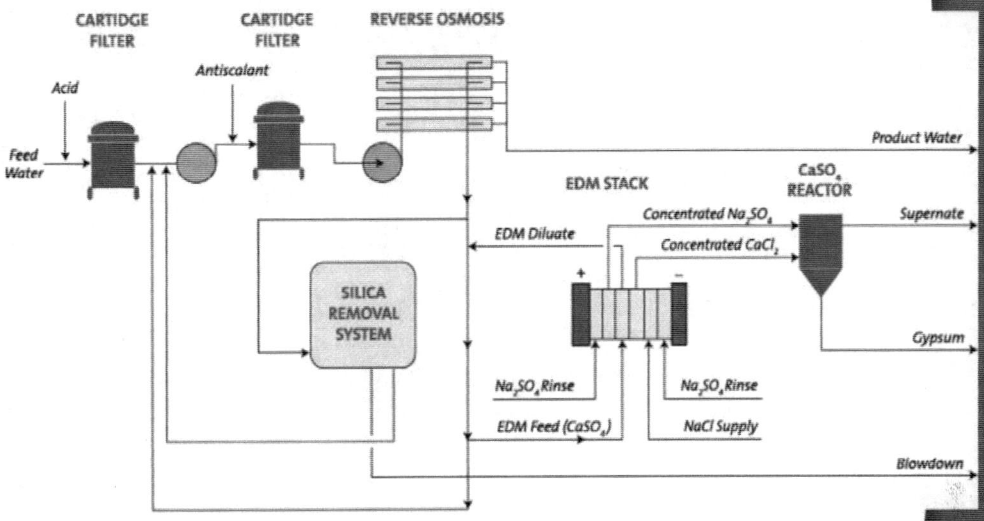

Treatment Options

- 4.0 MGD Feed Water Design Basis Comparing four different operating scenarios
  - ➢Conventional BWRO (75% Recovery) → **511 gpm** Evap Pond
  - ➢Chemical Softening + High Recovery RO (90% Recovery)
  - ➢ZDD with NF + EDM (98% Recovery)
  - ➢ZLD with RO + Evaporator + Crystallizer (>99% recovery)

Thermal Technologies

- Evaporators
  - Rising film
  - Falling film
  - Natural circulation
  - Forced circulation
- Crystallizers
  - Enhanced Forced Circulation
  - Forced circulation crystallization
  - Falling film crystallizers

CoLD^{TM} Process

Crystallization

of high solubility salts at

Low Temperature and

Deep Vacuum

U.S. Patent 8,052,763

Past Practice

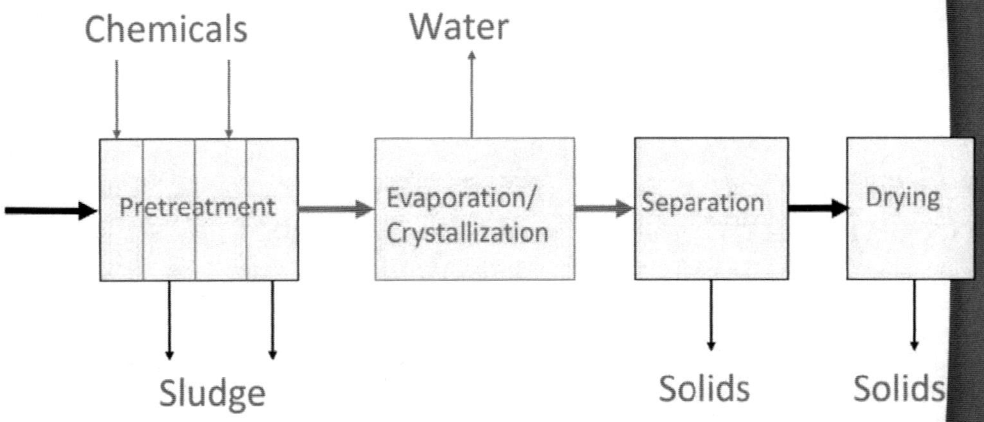

CoLD™ Process

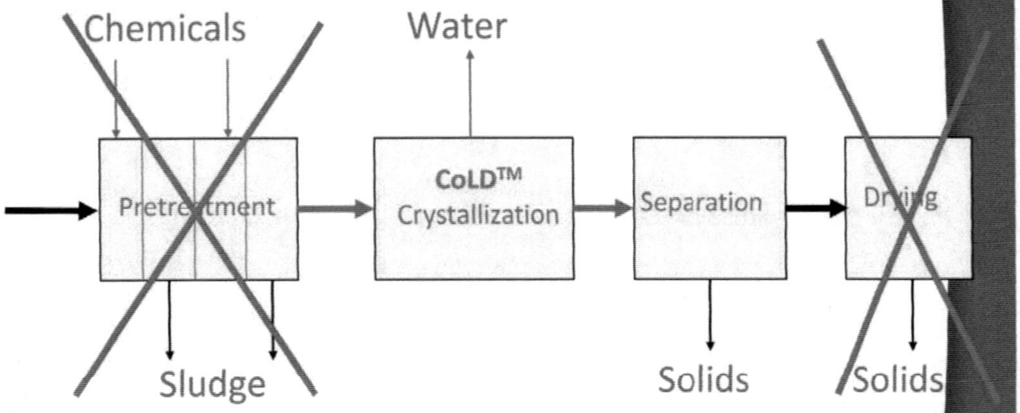

CoLD™ Advantages

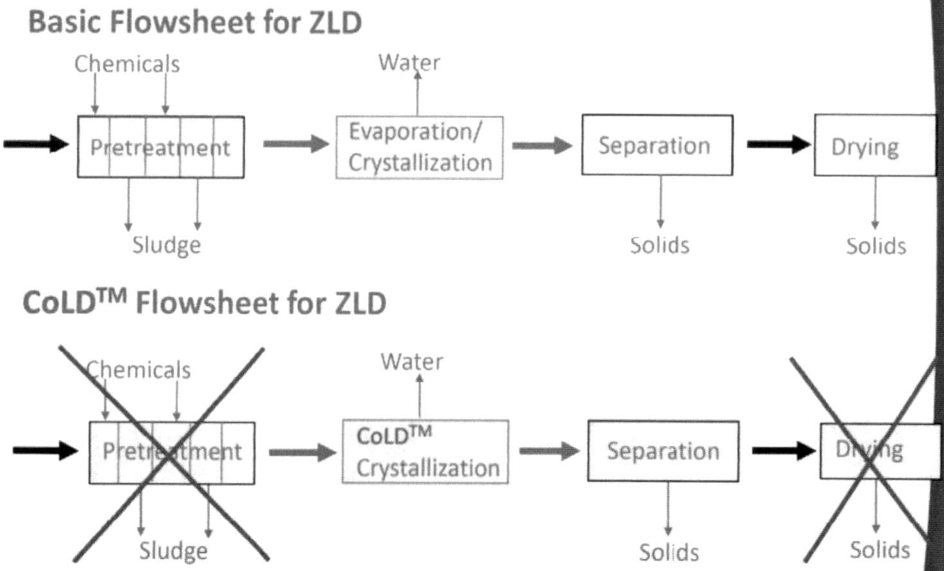

Advantages of CoLD™ Process

- No Chemicals, Less Sludge
- Lower Energy Requirement
- Simple System, Less Operator Attention
- Robust System
- Lower Capital Cost

ZLD System for IGCC Power Plant (Start-up 2012)

- 630 MW IGCC Power Facility
- Zero Liquid Discharge System by HPD Includes two CoLD™ Crystallizers
- Produces Clean Water for Discharge and Dry Salt for Landfill Disposal

ZLD System for IGCC Power Plant (Start-up 2012)

- 335 MW IGCC Power Facility
- Zero Liquid Discharge System by HPD Includes CoLD™ Crystallizer
- Produces Clean Water for Re-use and Dry Salt for Landfill Disposal

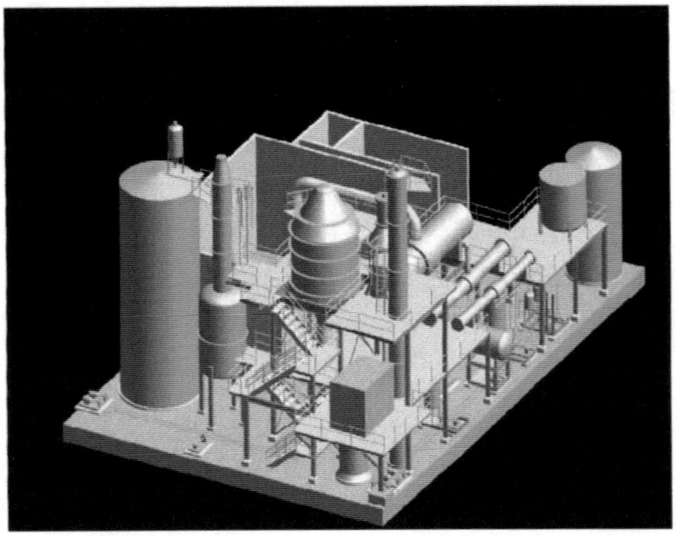

EW 60000 Spectra Energy Canada

- Evaporator
  - Double Effect Forced Circulation
  - Also Forced Circ MVR design

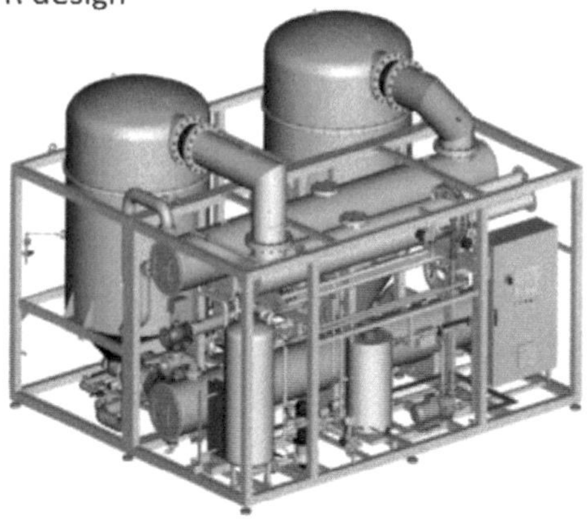

RW Type Concentrator Rio Tinto Mining

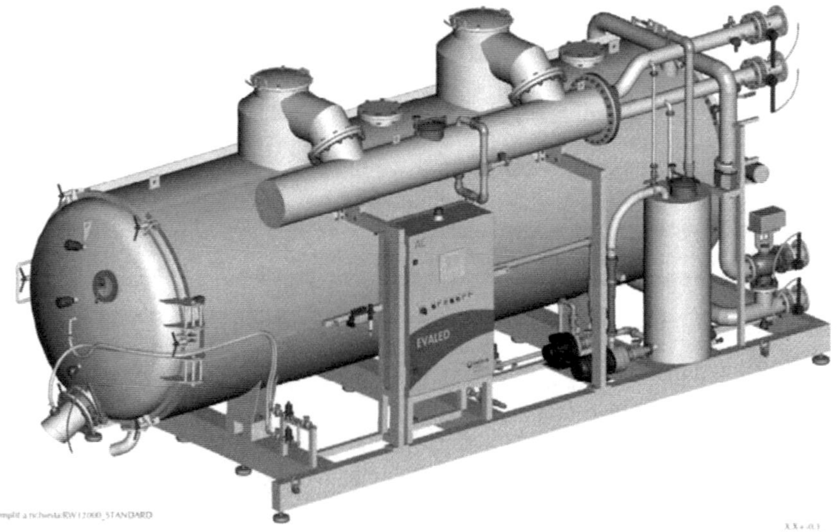

ACTIFLO® Turbo - Aquamove Trailer

Mobile Pilot Demonstration Units – Biological Treatment

NEOSEP® and AnoxKaldnes™ MBBR

6

Case Studies

- Summary
- Project Specifics

Weig, Mayen, Germany

- ❑ Before Upgrade: 300 m3/h
- ❑ WWTP Expansion to 500 m3/h
- ❑ Limited Space for Expansion of Secondary Clarifier

Start-up Year :	2006
Application :	Secondary Clarif. of AS for REUSE as PW (Plant Second. Clarifier Limiting Factor)
Market Segment :	Pulp & Paper
Total Capacity :	7 MLD
Number of Units :	1
Configuration :	ACTIFLO w/o Lamella; 30 m/h Rise Rate
Raw Water Source :	Activated Sludge Basin
Parameters :	TSS in/out: up to 6,500 / < 50 mg/l Limit

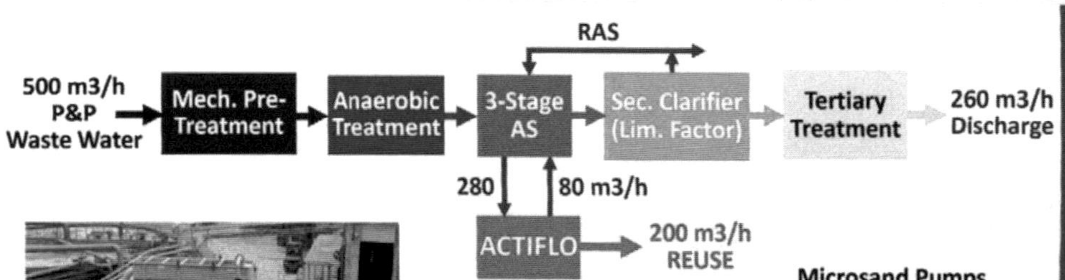

ACTIFLO Fitted in an extremely Tight Space

Weig Details

Microsand Pumps

Weig Paper Process Scheme before Upgrade

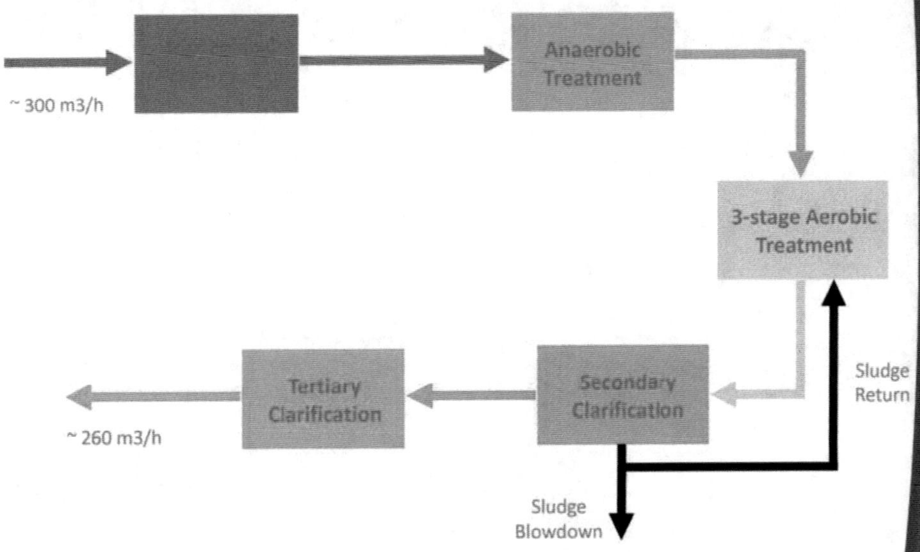

Weig Paper Final Process with Upgrade

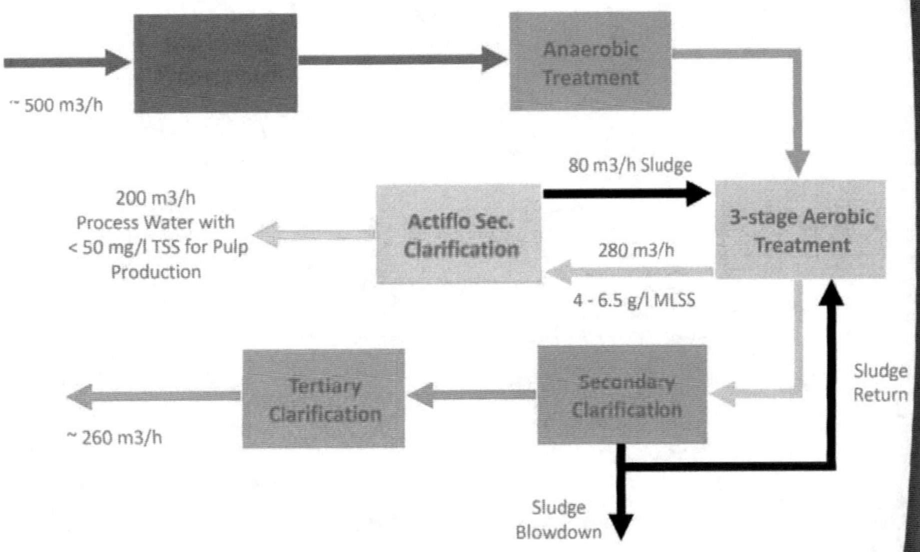

Weig Limited Space to Upgrade System

Actiflo Installed Upgrading Plant Performance

Idemitsu Kosan Chita, Japan

Start-up Year :	2008
Application :	REUSE WW for Boiler Feed
Market Segment :	Oil & Gas Refining/Petrochemical
Total Capacity :	2.9 MGD (11 MLD) + 13.7 gpm AFM
Number of Units :	Single Actiflo Train
Configuration :	ACTIFLO Mini for Sludge Thickening
Raw Water Source :	Industrial Waste Water
Parameters :	Algae > 90% Removal, < 2 NTU Limit

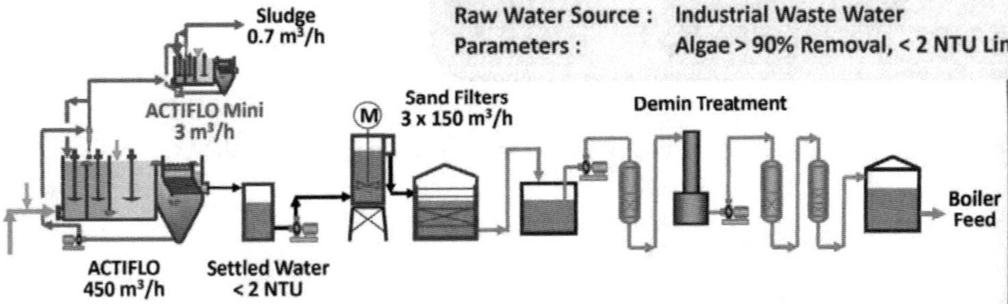

Idemitsu Kosan, Chita, Japan - 1 x 450 m^3/h (2008)

- DESIGN CONDITIONS:
  - ▶Remove Turbidity & Algae *(Melosira ~3,500 cells/ml)*
  - ▶Inlet Raw Water Turbidity; (Design < 15 NTU); (Actual 1.5 - 6 NTU)
  - ▶Treated Turbidity: < 2 NTU
- OPERATING RESULTS:
  - ▶Minimize Liquid Sludge Flow to < 4.4 gpm (< 1 m3/h)
  - ▶Treated TSS: Algae > 90% removal
  - ▶ Treated Turbidity: 2 NTU
- TREATMENT FEATURES:
  - ▶ Liquid Sludge from the Main Actiflo is treated with a smaller Actiflo (aka Mini ActifloMini to reduce the volume by % 40
  - ▶ The Mini Actiflo is mounted above the main Actiflo unit to minimize the footprint to (60' LG X 12'W)
  - ▶ Able to remove a majority of the Algae and produce 2 NTU water

LG Siltron P3 - 1 x 264 gpm (60 m3/h) Gumi, S. Korea

Start-up Year: 2005

Industry: Semiconductor – Silicon Wafer Manufacture

Application: High TSS water from silicon cutting & polishing process

Equipment: Actiflo + Hydrotech Discfilter (Actidisk), MF, RO

Goal: $SDI_{15} < 3$; Turbidity < 1 NTU

Total Capacity: 264 gpm (60 m3/h) (1.4 MLD)

No. of Units: 1 Train

Water Source: Industrial Waste Water

Parameters : Please refer to Process Schematic for additonal details

Advantages:
- Packaged design providing a compact footprint
- Start-Stop operation to match wafer manufacturing demand
- High Automated with little or no operator attention

LG Siltron P3, Process Schematic with Operating Results

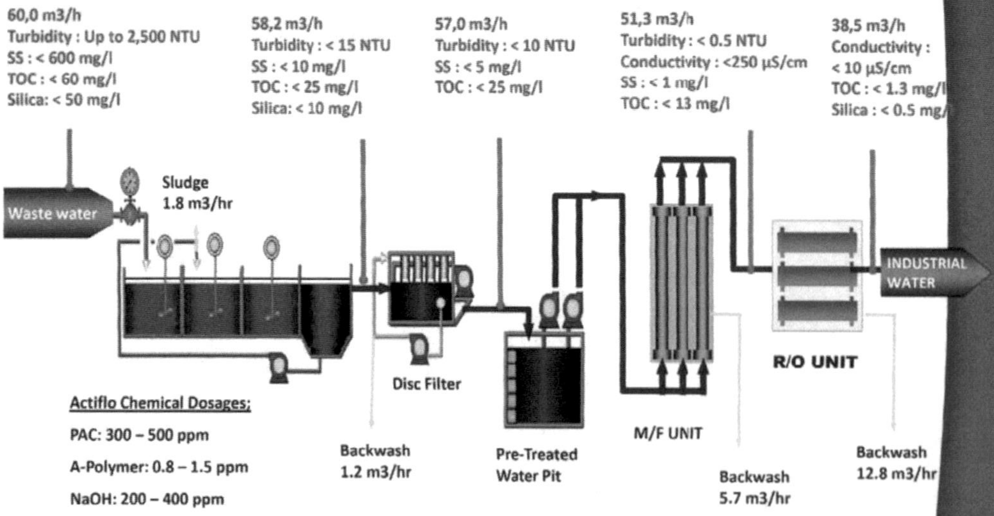

LG Siltron P3 - Actidisk Equipment Arrangement

Chevron San Ardo Facility Produced Water Case Study

- OPUS Technology Applied
  - Treat PW for REUSE:
    - →Surface Water Discharge
    - →OTSG Make-Up
  - System Capacity
    - →2,100 gpm
  - Recovery Rates :
    - →System Recovery > 75%; RO > 81%
  - RO CIP Requirements
    - →Once 3 to 4 months; No Standby RO, > 99% Availability

Chevron San Ardo Facility
Produced Water Case Study - Treatment

CONSTITUENT (ppm)	PRODUCED WATER	[1] SURFACE DISCHARGE QUALITY SPECIFICATION	FINAL TREATED QUALITY
TDS	7,000	600	120
SODIUM	2,300	100	43
CHLORIDE	3,400	150	11
SULPHATE	133	150	120
NITRATE	10.0	5.0	N.D.
BORON	26.0	0.64	0.24
pH	7.5	6.5 ~ 8.4	7.0

[2] For Once Through Steam Generation (OTSG) Total Hardness <2 ppm as $CaCO_3$

Produced Water System Jobsite
Chevron San Ardo Facility

Darling Quarter Commonwealth Bank Place Sidney, Australia

- Commercial office and leisure space redevelopment
- Located in Darling Harbour next to the Sydney CBD
- Green Star rating: 6 Star/World Leader

greenstar

BUILDING DESIGN REDUCES:
- CO_2 emissions by 2,500 tonnes per year
- Potable water use by 92%
- Very small footprint available for the RWP in the basement of the building (2475 ft2) (230 m2)

Darling Quarter REUSE Innovative Technology Combination

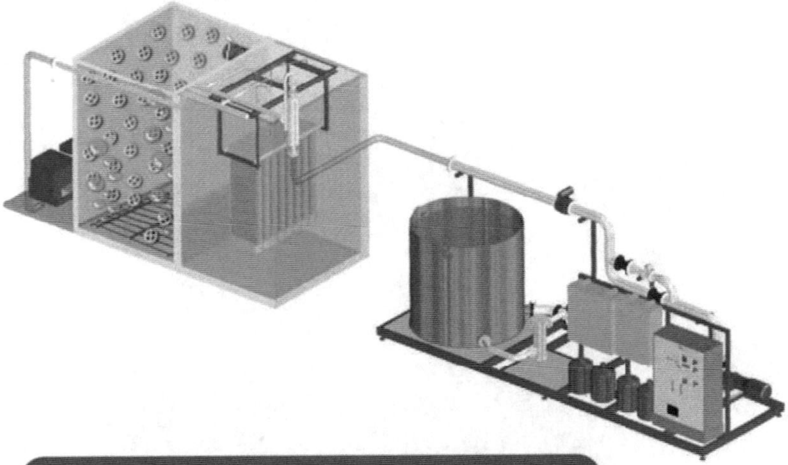

MBBR + UF

Darling Quarter REUSE
Simplified Block Flow

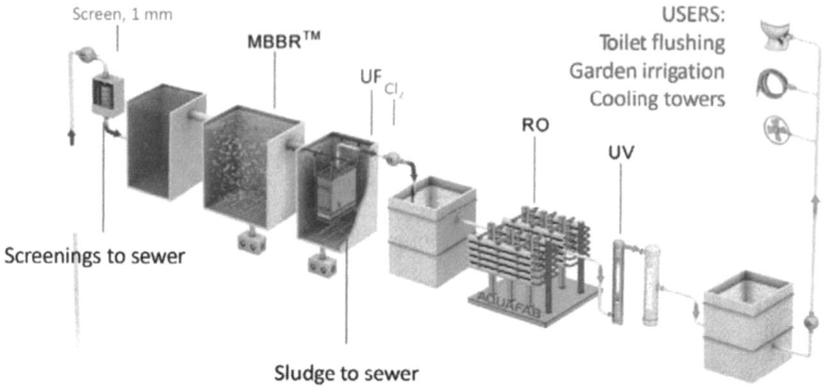

7.9 MM gals (60 MM liters) of drinking water are saved each year

10

Darling Quarter REUSE
Equipment Footprint

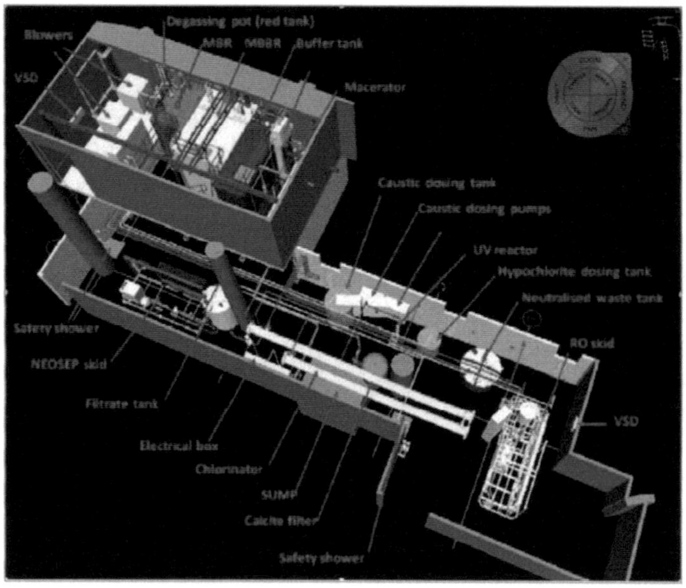

9

Treated Water Quality exceeds Australian Recycled Water Guidelines

PARAMETERS	WATER QUALITY ACHIEVED
BOD5	<5 mg/L
TSS	<5 mg/L
pH	6.0-9.0
Turbidity	<0.2 NTU
Conductivity	20 – 50 μS/cm
E. Coli	<1cfu/100mL
Coliphages	<1pfu/100mL
Clostridia	<1cfu/100mL
Validated virus reduction	6.6 log reduction
Validated bacteria reduction	12.1 log reduction
Validated protozoa reduction	8.1 log reduction

MBBR+UF Performance Compared to MBR

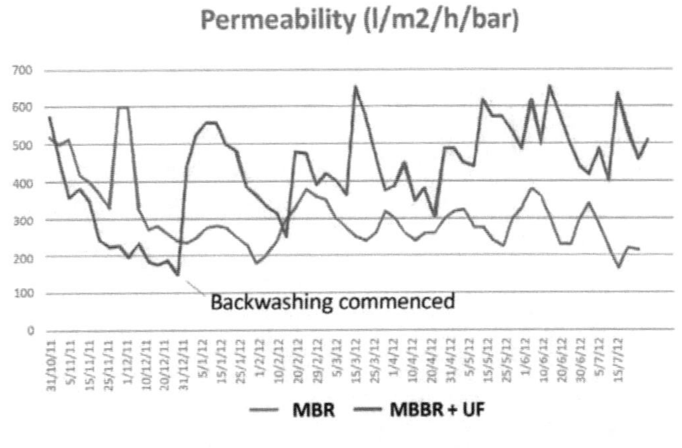

- Higher permeability and longer cleaning intervals (4 times less chemical used) for MBBR + UF compared to MBR

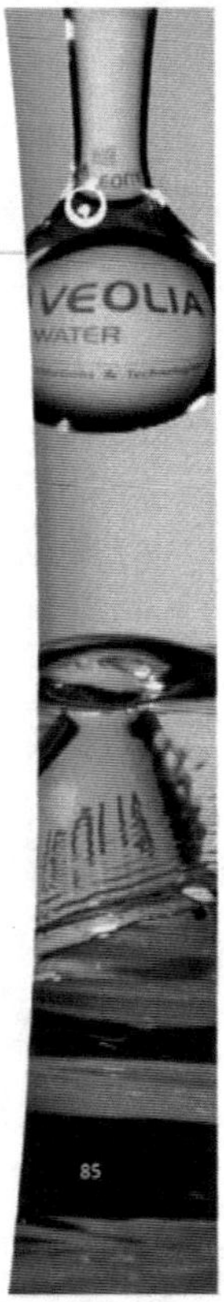

Recycled Water Plant Performance

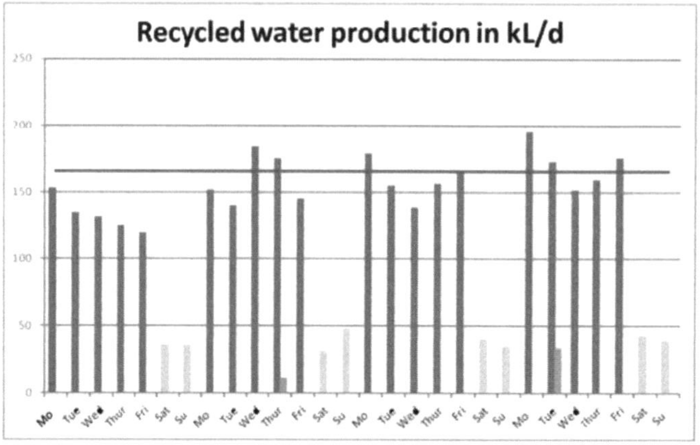

- Potable water top off when demand is too high; > 98% utilization using produced recycled water
- Darling Quarter potable water consumption is 0.18 kL/m2/year (more than 4X below best practice guidelines : 0.77kL/m2/Year)

14

Darling Quarter Engineering Excellence Awards Winner Sydney 2012

- Building and Structures category

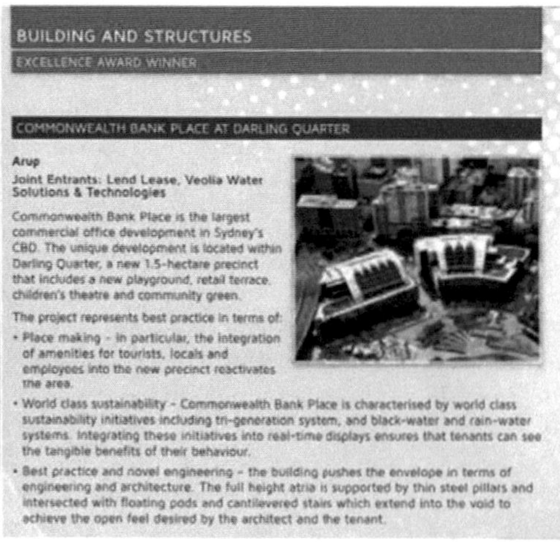

BUILDING AND STRUCTURES

EXCELLENCE AWARD WINNER

COMMONWEALTH BANK PLACE AT DARLING QUARTER

Arup

Joint Entrants: Lend Lease, Veolia Water Solutions & Technologies

Commonwealth Bank Place is the largest commercial office development in Sydney's CBD. The unique development is located within Darling Quarter, a new 1.5-hectare precinct that includes a new playground, retail terrace, children's theatre and community green.

The project represents best practice in terms of:

- Place making - in particular, the integration of amenities for tourists, locals and employees into the new precinct reactivates the area.
- World class sustainability - Commonwealth Bank Place is characterised by world class sustainability initiatives including tri-generation system, and black-water and rain-water systems. Integrating these initiatives into real-time displays ensures that tenants can see the tangible benefits of their behaviour.
- Best practice and novel engineering - the building pushes the envelope in terms of engineering and architecture. The full height atria is supported by thin steel pillars and intersected with floating pods and cantilevered stairs which extend into the void to achieve the open feel desired by the architect and the tenant.

22

SHELL Pearl GTL Complex, Qatar
Waste Water / REUSE / ZLD

REUSE

Client Overview & Requirements

The Pearl GTL project (2011) comprises the development of upstream gas production facilities, as well as an onshore GTL plant to produce:
140,000 bpd of GTL products
120,000 bpd of associated condensate
LPG (Ethane) 320,000 bpd

Project Description

- Effluent Treatment Plant
- Treatment of the cooling water blow-down for reuse in the process
- Management and optimization of water cycle targeting Zero-Liquid Discharge target
- **2 X 4400 gpm; all of the facility WW is treated to supply Process and Boiler Feed Water**

Shell Project Investment ...$19B; $500MM for Water Treatment REUSE

Technologies & Solutions

- Several unit processes including UF & RO
- Actiflo (100% effluent recycle)
- Sludge treatment using BC+C to minimize volume production of dewatered sludge and salt

Technical Data

- Removal of solids, oil, dissolved organics, salts and pathogens

SHELL Qatar, Pearl GTL complex, Qatar

Technologies and Technical data

- <u>**Cooling Tower Blowdown:**</u>
- **Hollow fiber SUF** PVDF membranes built with interchangeability for future maintenance; recovery rate > 90%
- **Single-pass RO** units with sea water membranes. The recovery rate is > 85% and the operating pressure = 16 bars.
- <u>**Effluent Treatment (ETP) and Water Reuse:**</u>
- Influent contains: Oil, organic acids, alkalinity
- CPI units for oil/water separation
- **2 Biotreater trains** using activated sludge
- **Hollow fiber SUF** PVDF membranes built with interchangeability; to meet future maintenance; recovery rate > 85%
- **3 Double-Pass RO** units for TDS removal, including CIP units. Trains 1 and 2 have a recovery rate > 80%; the third common train has a recovery > 30%. Operating pressure is 19-20 bars.

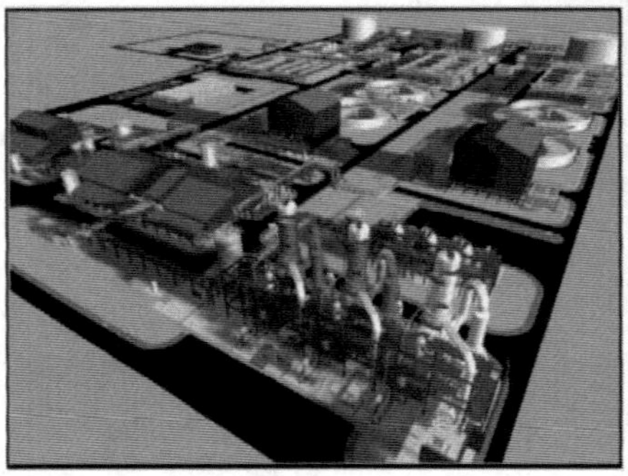

- Brine from RO is treated by **Evaporation** and **Crystallization** to achieve zero liquid discharge.
- Actiflo™ clarification
- Bio-sludge dewatering through centrifugation before composting.

SHELL Water Re-use with Zero Liquid Discharge Gas to Liquids (GTL) SHELL, Qatar

Overall Scheme of Main Units and Flows:

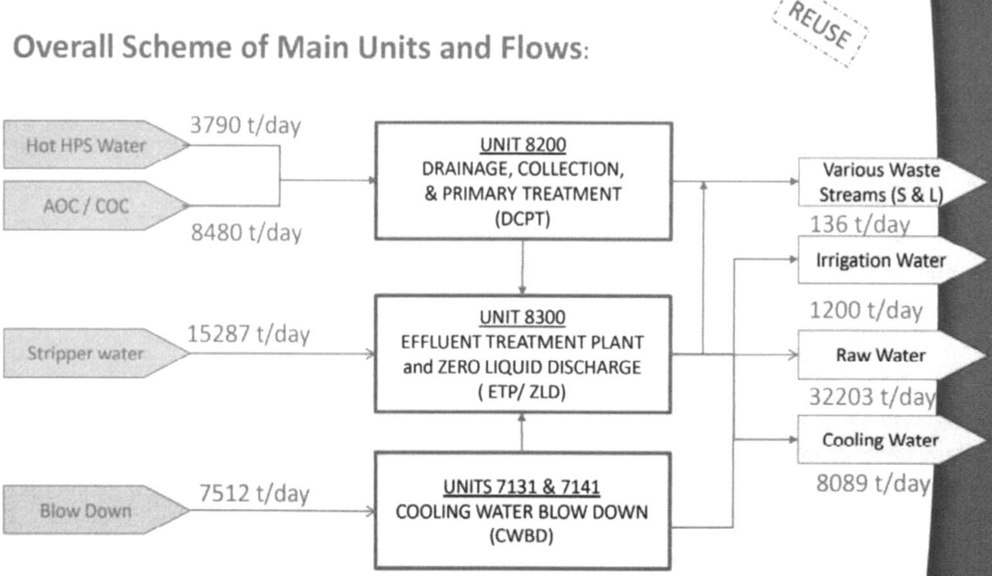

SHELL PEARL GTL, QATAR - Effluent Plant

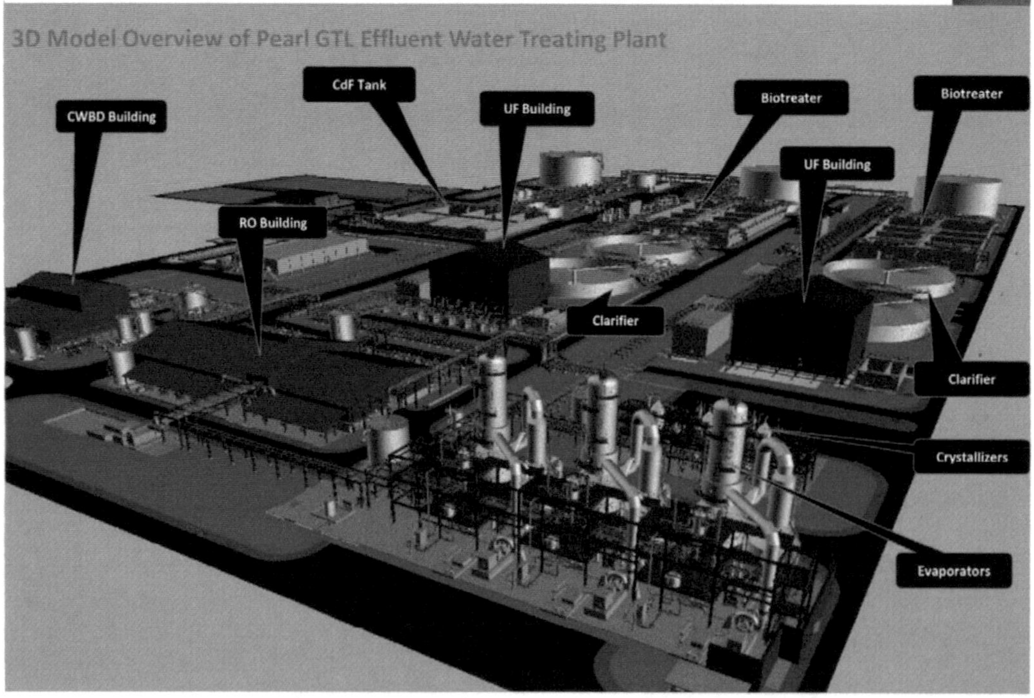

FFU, Biotreaters & Clarifiers

ZLD for Shell Pearl GTL complex – Qatar - EPC contract delivered in 2011

SHELL Qatar Pearl GTL Complex Qatar

Submerged Ultrafiltration

Reverse Osmosis

Evaporation

Biotreater

SASOL Landlord, South Africa
REUSE

sasol reaching new frontiers

REUSE

Client Overview & Requirements

- Established in 1950 by South African government, Sasol is a global energy company with operations in chemicals, fuels and gas, and has now 50 entities w/ 30,000 employees.
- Headquartered in Johannesburg, South Africa engaged in the commercial production and marketing of chemicals and liquid fuels; with a growing interest in oil and gas exploration.
- Waste Water being reused for Sasol's fuels enhancement and polymers expansion project.

Project Description

- Cooling towers blowdown reuse through construction of water treatment plant and water purification installations

Technologies & Solutions

- 2 Multiflo® clarifiers, a patented process, using lamella plates
- 3 treatment trains consisting of:
  - Submerged membrane.
  - Ion exchange softeners
  - Reverse osmosis & MB demin system

Scope of Supply

- Design, construction and commissioning

Technical Data

- See treatment line details

SASOL Landlord, South Africa
REUSE

sasol reaching new frontiers

REUSE

Treatment line details

- Clarified Water
  - MULTIFLO 300 feeding the demineralization unit, chill tower of oxygen plant, process cooling water make up, and also desalination plant with reduced flow
- Cold Lime Softener
  - MULTIFLO 300 with lime and soda-ash to reduce TOC, hardness, silica
  - Also treating CTBD for (full reuse) and
- Submersible Membrane
  - Pre-treatment through T- type strainers with 1 mm screen opening
    - → Control of pH, Turbidity, Conductivity, Hardness
    - → UF can also be fed directly, with clarified water or even CTBD water
  - UF Membrane system (7075 m^2 per train, 36 $l/m^2.h$, outlet with SDI< 3)

- IX Exchange
- RO
- Mix Bed
- Flow-rate : 4580gpm

PETROBRAS RNEST
Pernambuco , Brazil

REUSE

- CLIENT: bpd) with Plant — Petrobras new refinery (200.000 a Water/REUSE Water Treatment 12.7 mgd/8800 gpm (2000 m3/h)
- MARKET: HPI/CPI
- START DATE: 03.2009
- FINISH DATE: 12.2012
- EXECUTION: Erection + — Engineering + Equipment + Civil + Commissioning + Assisted Operation
- MAIN PROCESS: Floto-filtration >>>Demineralization>>> Sludge Treatment >>>Dewatering (Centrifuge)>>> GAC Filtration>>>EDR Treatment + Blowdown Treatment (using Multiflo)

PETROBRAS RNEST
Pernambuco , Brazil

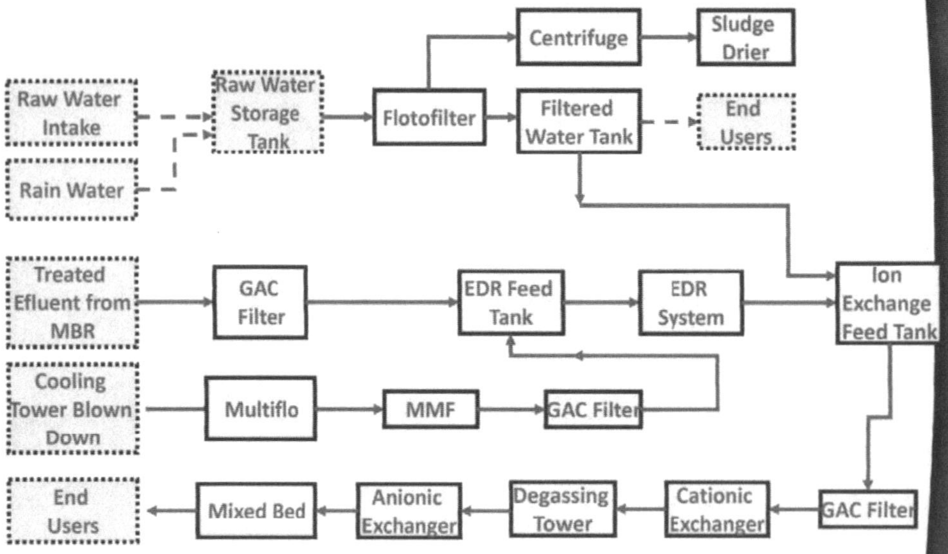

PETROBRAS RNEST – Return Condensate Treatment 3.5 mgd/2467 gpm (560 t/h)

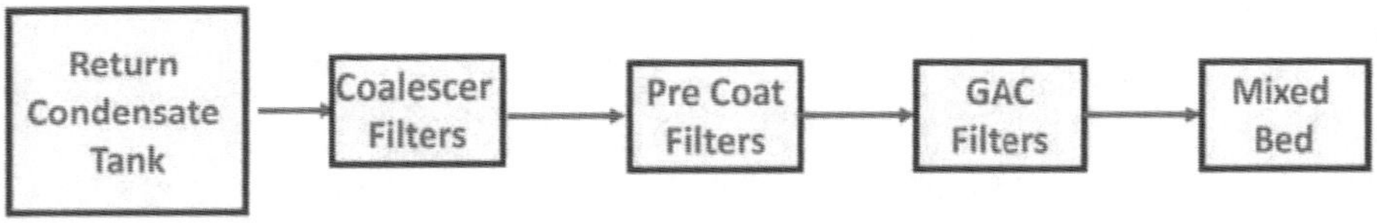

PETROBRAS RNEST
Pernambuco , Brazil

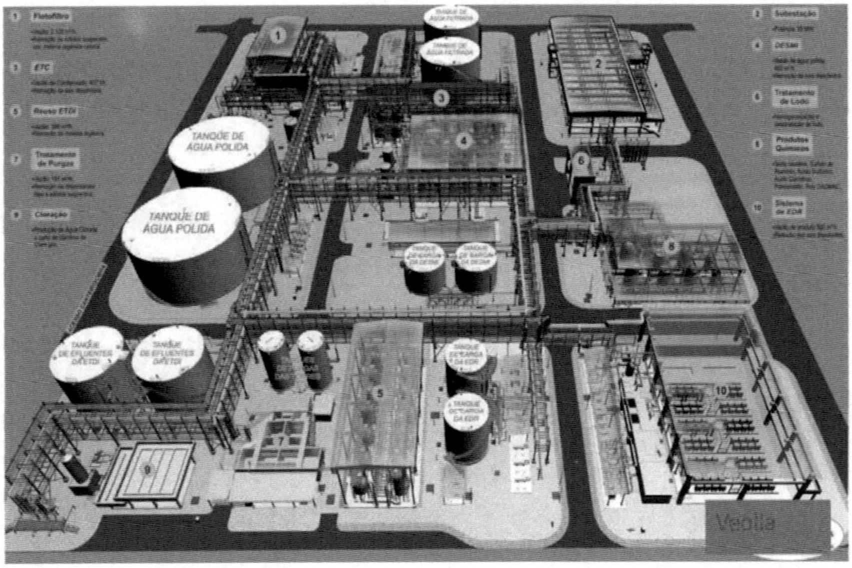

PETROBRAS RNEST
Pernambuco , Brazil

PETROBRAS RNEST
Pernambuco , Brazil

CONSOL Energy Mine Water Treatment
The Issue – Mine Water Discharge

- West Virginia Department of Environmental Protection new regulations on chlorides in discharges to receiving streams
- Three mines discharge conventionally treated mine drainage from six different locations
- Current treatment for pH, metals, and TSS removal at the points of discharge does not address Chloride removal
- Risk associated with the disposal of treatment residuals

CONSOL ENERGY
AMERICA'S ENERGY STARTS HERE

CONSOL Energy Mine Water Treatment
The Solution – Zero Liquid Waste

- After much review Consol Energy decided to move forward with a centralized treatment facility that creates Zero Liquid Waste (ZLW)
- The ZLW concept differs from ZLD in that a large quantity of water is discharged, however this water is compliant with environmental discharge requirements
- The water that is discharged is low in TDS and chlorides improving the overall quality of the river allow users located downstream to use this water
- Waste residuals are condensed into a solid that is properly disposed in an on-site landfill
- No liquid waste from the water treatment operation leaves the site

CONSOL ENERGY
AMERICA'S ENERGY STARTS HERE

CONSOL Energy Water Treatment - Mine Locations

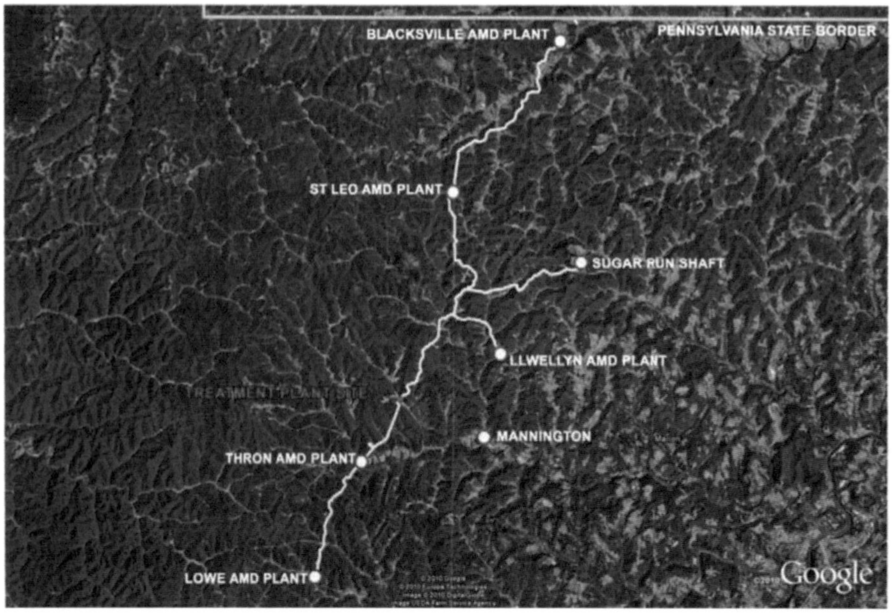

CONSOL Energy Water Treatment - Project Challenges

- Substantial capital and human resources required to meet the requirements for:
  - Project Execution
  - Facility construction (tight schedule)
  - Commissioning & Operations

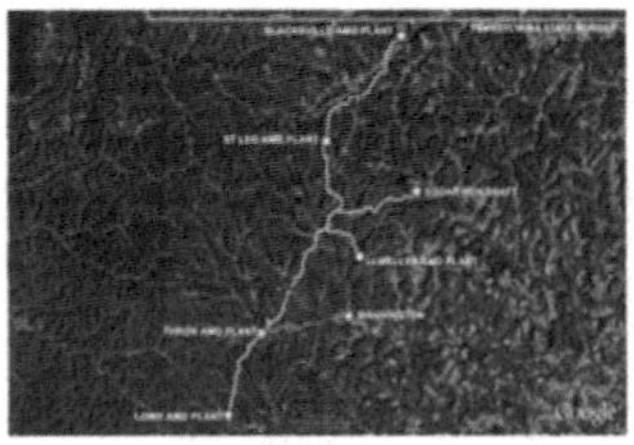

CONSOL Energy - Design Criteria

Influent Water Quality

Parameters	Maximum
Design Flow, GPM	3500 max (1750 min)
pH, S.U.	5 – 10
Temperature, deg F	38 – 85
Chlorides, mg/L	1,500
Sulfates, mg/L	5,500
Iron, mg/L	150
Manganese, mg/L	2
Aluminum, mg/L	8
TDS, mg/L	10,000
TSS, mg/L	150

Key Effluent Requirements

Parameters	Maximum Effluent Concentration
Chlorides, mg/L	< 218
TDS, mg/L	<150*
pH, S.U.	6 to 9

CONSOL Energy - Facility Overview

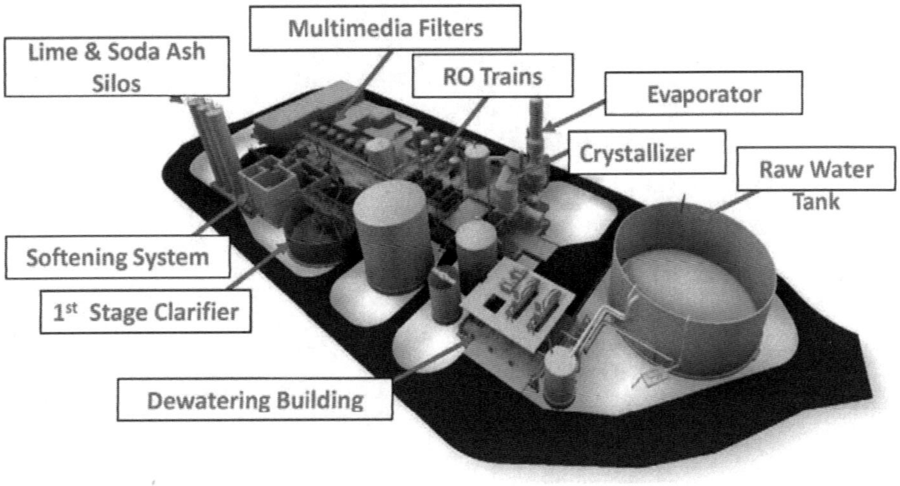

Consol Energy - Reverse Osmosis System

Evaporator Design – Falling Film

Brine Management System

Consol Energy - Treatment Plant Site

Photograph from mid-July 2012

LG Empire Upstate NY Power Project

- Commercial operation began in September 2010
- New York State's voluntary Brownfield Cleanup Program (reinvestment of resources in the environmental remediation and redeployment of existing non-productive sites)
- Use of municipal effluent (gray water) rather than fresh (groundwater) water resource

114

LG Empire NY Power Project – Water Quality

Ions	Gray Water	Sand Filter	Cl_2 Contact	Demin	Demin (treated)
Turbidity NTU (avg./max/peak)*	3.8 / 7 / ---	2 / 5 / 10			< 1
TSS (ppm) (avg./max/peak)*	8 / 22 / 45	7 peak			< 1
Conductivity (μS)	747			< 0.1	**0.05 ppb**
TDS (ppm)	780				
Silica (mg/l as SiO_2)	7.2			< 10 ppb	**3.1**
Sodium (mg/l as ion)	89			< 10 ppb	**0.4 ppb**
TOC (ppm)	5.7			< 100 ppb	**35/48 ppb**
Chloride (mg/l as ion)	144			< 10 ppb	**0.003 ppb**
Sulfate (mg/l as ion)	38			< 10 ppb	**< 0.005 ppb**
DOC (ppm)	5.0				
CA Title 22 Sec. 60301.230 (a) 1 & (b)			Yes		

*Maximum allowable ≤ 5% of 24 hour period. Peak not to be exceeded at any time.

115

7

Water Impact Index

- True Cost of Water
- WIIX
- Case Example

Journey Towards Sustainable Growth

CARBON FOOTPRINT → WATER IMPACT INDEX → ENVIRONMENTAL FOOTPRINT → GREEN FOOTPRINT

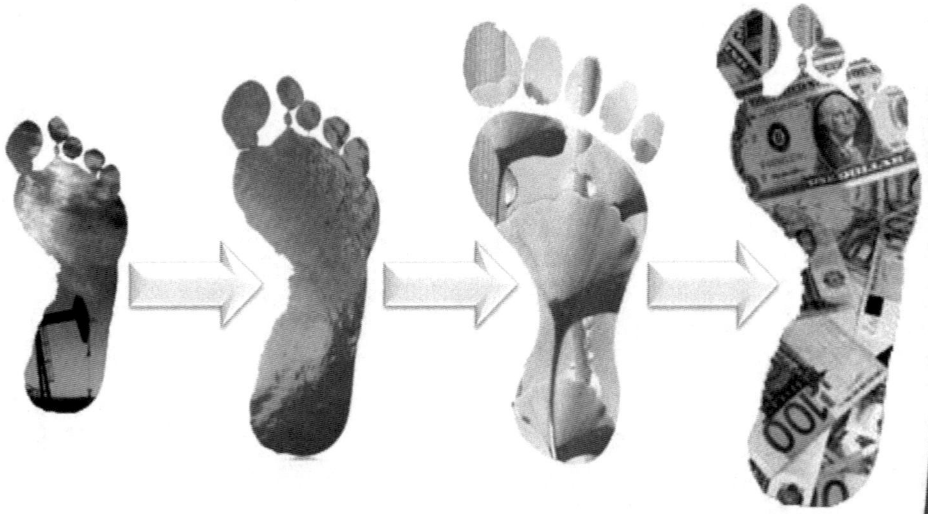

True Cost of Water

The Environmental Footprint
"a decision-making tool for sustainability"

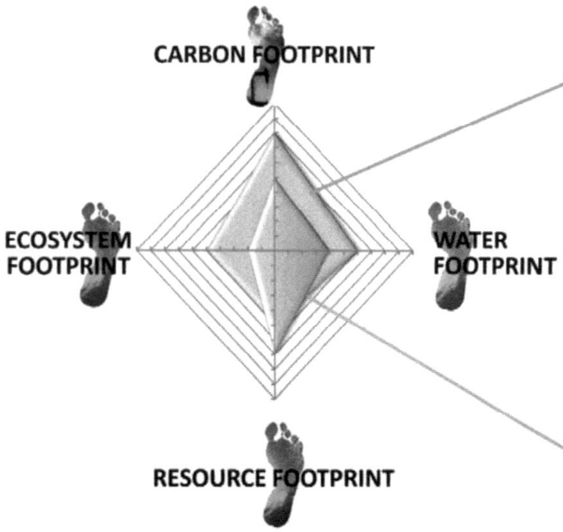

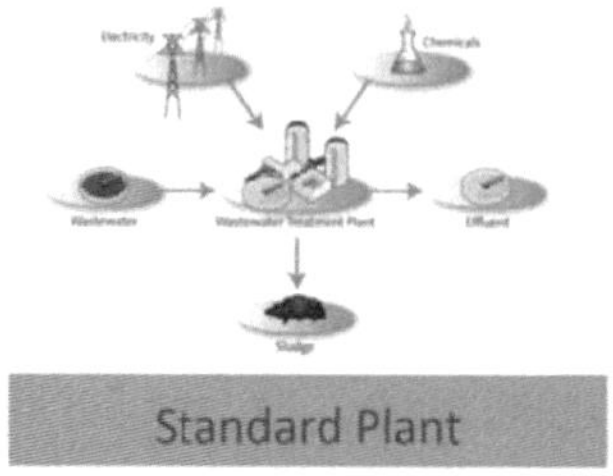

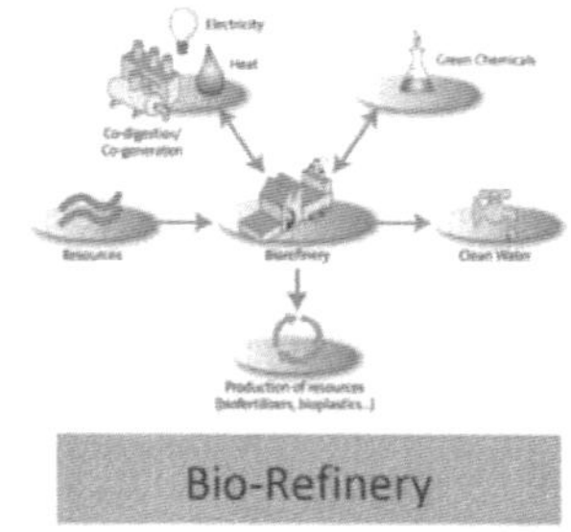

Veolia Environnement's Vision

Become *"the benchmark for sustainable growth"*

Antoine FREROT

Three main challenges :

-> Decarbonate (Reduce CO_2 Emissions)

-> Dehydrate (Reduce Water Impact Index)

-> Dematerialize (Reduce Resource Consumption)

Fills a critical need by defining the standards used to measure and make decisions

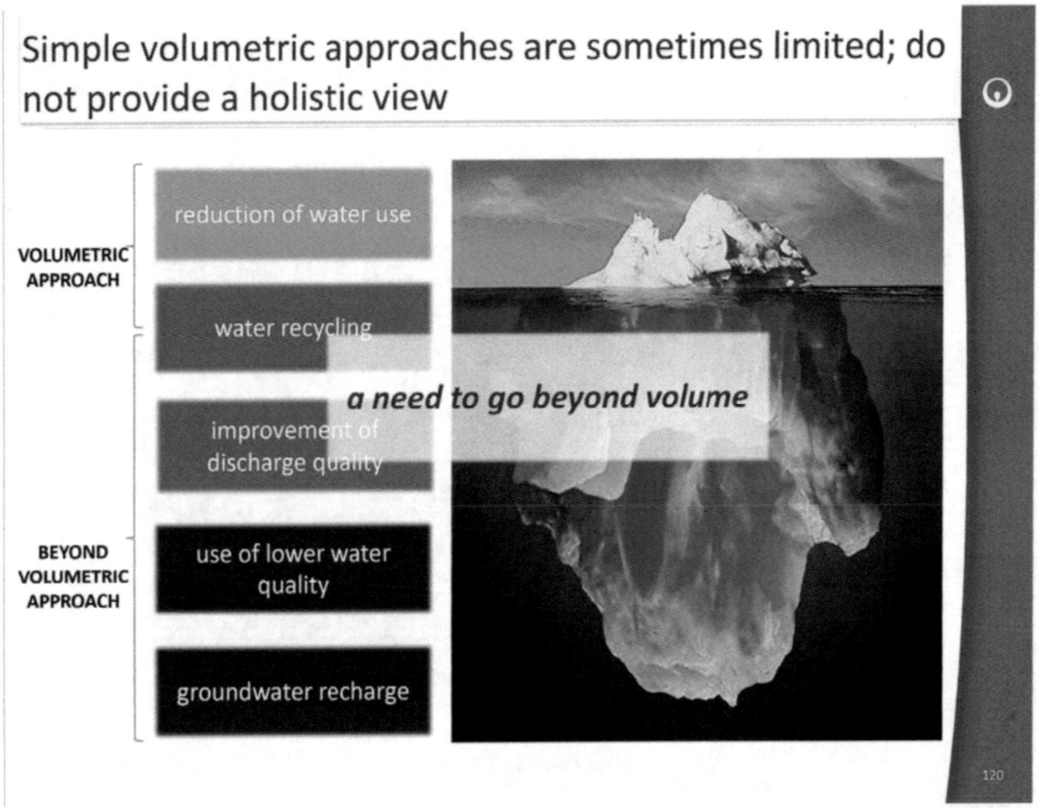
Simple volumetric approaches are sometimes limited; do not provide a holistic view
VOLUMETRIC APPROACH
BEYOND VOLUMETRIC APPROACH
reduction of water use
water recycling
improvement of discharge quality
use of lower water quality
groundwater recharge
a need to go beyond volume
120

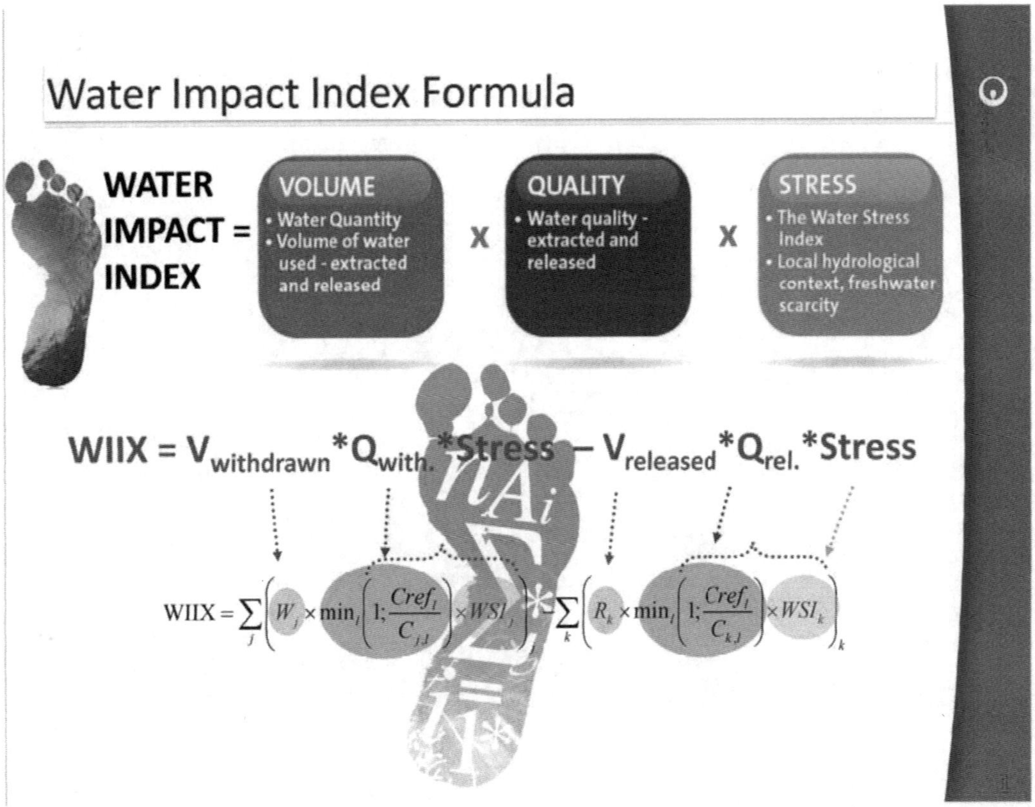
Water Impact Index Formula
WATER IMPACT = INDEX
VOLUME
• Water Quantity
• Volume of water used - extracted and released
x
QUALITY
• Water quality - extracted and released
x
STRESS
• The Water Stress Index
• Local hydrological context, freshwater scarcity
WIIX = V_withdrawn *Q_with. *Stress – V_released *Q_rel. *Stress
WIIX = Σ_j (W_j × min_l (1; Cref_l / C_j,l) × WSI_j)_j − Σ_k (R_k × min_l (1; Cref_l / C_k,l) × WSI_k)_k

Water Stress Index

- The WSI takes into account:
  - Water Usage
  - Availability
  - Stored Water (water in stock)
- An independent index
  - Created by Pfister (Zürich) in (2009)

0 < WSI < 1

No Stress — High Stress

WSI=0.996 -> High Stressed Area

No Stress

Available on Google Earth after downloading the link below
http://www.ifu.ethz.ch/staff/stpfiste/Impact_factors_LCA_pfister_et_al.kmz

123

Water Stress Index

0 < WSI < 1

High Stress

No Stress

Available on Google Earth after downloading the link below
http://www.ifu.ethz.ch/staff/stpfiste/Impact_factors_LCA_pfister_et_al.kmz

124

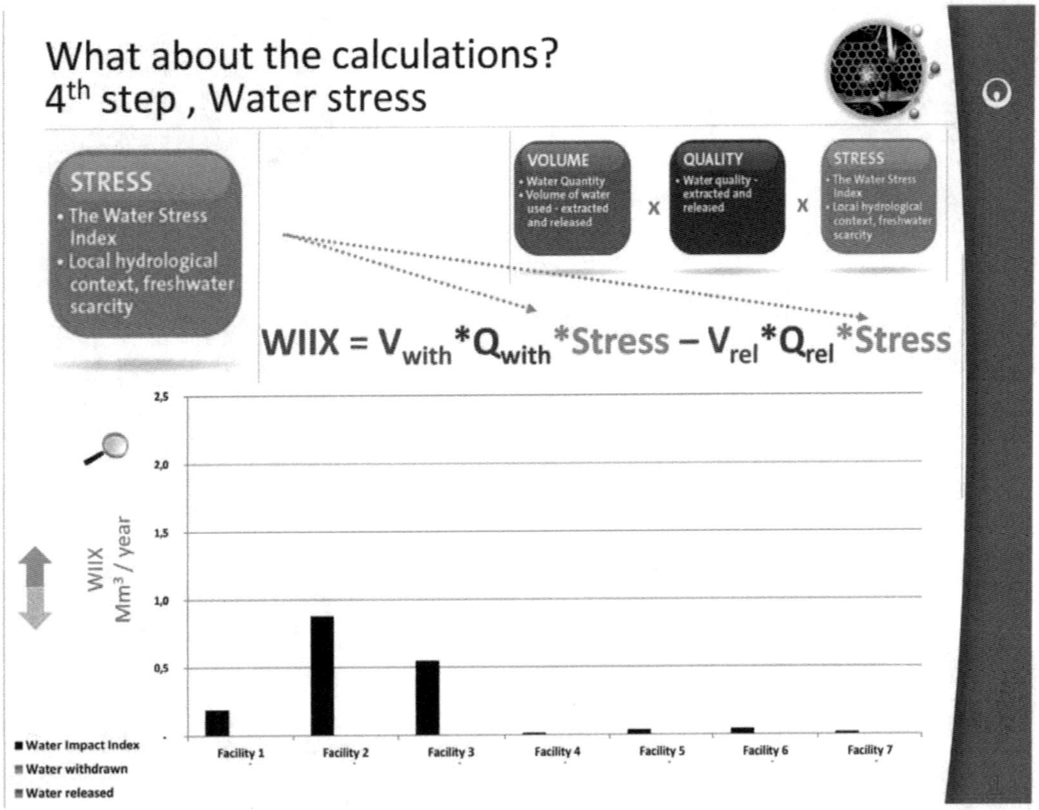
What about the calculations?
4th step , Water stress
STRESS
• The Water Stress Index
• Local hydrological context, freshwater scarcity
VOLUME
• Water Quantity
• Volume of water used - extracted and released
X
QUALITY
• Water quality - extracted and released
X
STRESS
• The Water Stress Index
• Local hydrological context, freshwater scarcity
$WIIX = V_{with}*Q_{with}*Stress – V_{rel}*Q_{rel}*Stress$
WIIX
Mm³ / year
2,5
2,0
1,5
1,0
0,5
Facility 1
Facility 2
Facility 3
Facility 4
Facility 5
Facility 6
Facility 7
Water Impact Index
Water withdrawn
Water released

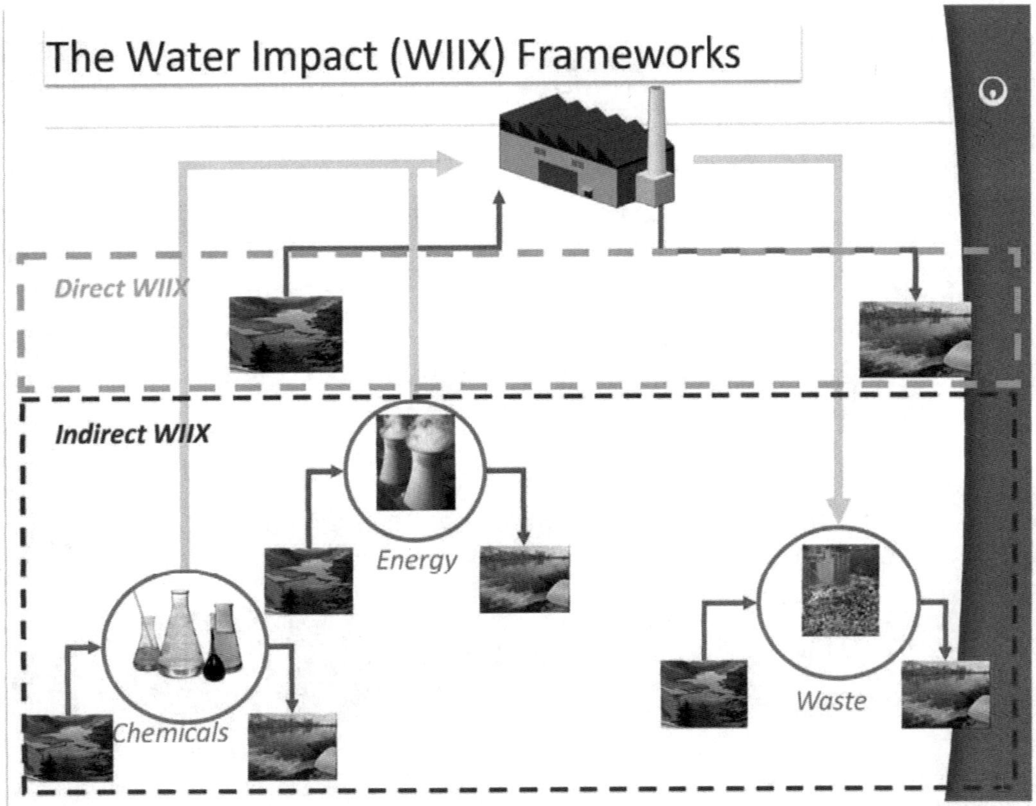
The Water Impact (WIIX) Frameworks
Direct WIIX
Indirect WIIX
Energy
Chemicals
Waste

8

Close

- Summary
- Q & A

Conclusion

- Waste Water is a **valuable resource**
  - Can be used to drive sustainable growth
- Knowledge, integration, and technologies can be used to maximize the value of incorporating REUSE
- Investment in these technologies and a strong commitment to water reuse will:
  - Provide for the conservation of local fresh water resources
  - Protect the client's assets, performance and capital investment

This approach and mindset not only for ***sustainable water management***, but also for ***environmental***

128

Conclusion

- REUSE
  - The culmination of what we learned from across industry sectors and engineering disciplines covering municipal and industrial
- Paradigm Shift
  - From Water as a waste >>> Waste Water as a resource
- Not Easy, Not Quick, Not Free, Not "Not Needed"
- Believe it, Embrace it, Promote it

NO REUSE?

Questions, Discussion & Follow-Up

THANK YOU

Larry Gurnari
937.238.3965
larry.gurnari@veoliawater.com

Accomplishments and Committee Produced Publications

- On January 28-29, 2013, the Research Committee on Water Management Technology conducted the ASME Industrial Desalination Workshop: Best Practices and Future Directions. A workshop report will be issued in late Spring. Presentations included:

  o Workshop Goals and Background – Michael Tinkleman, ASME

  o Introduction to Industrial Desalination – Jane Kucera, Nalco

  o Overview of Desal Water Applications and Opportunities in Industrial Facilities – Loraine Huchler, MarTech Systems

  o An Overview of Industrial Desalination Technologies – Shahid Chaudhry, California Energy Commission

  o Industrial User's Perspective on Applications of Desalination to Meet Water Needs – Issues and Lessons Learned – Doug Hubbard, AEP

- In partnership with the Water Environment Federation (WEF), the Research Committee on Water Management Technology conducted the **Municipal Wastewater Reuse by Electric Utilities: Best Practices and Future Directions Workshop** on May 21-22, 2012. This workshop was conducted as part of the committee's efforts to develop industry specific workshops to promote and capture best management practices. A report of the workshop results was issued in September 2012. **Download the workshop report.**

- Prior to the formation of the Research Committee, several of the current members help to organize and participated in the **Water Management Technology Vision and Roadmap Workshop** held in May 2007. The workshop served as a starting point for dialogue among ASME, industry, government, and academia in order to guide future ASME efforts to develop value-added products and services in the water management area that benefit both ASME members and the nation at large. The workshop was by invitation only with 35 participants from ASME and 24 organizations across the U.S. **Download the final Roadmap report.**

- In May 2009, the Research Committee conducted a **Water Management Technology Best Management Practices Workshop** for the Process Industries. The intent of the workshop was to capture current best management practices (BMPs) and technology solutions in industrial water management as well as to initiate the development of a new ASME approach to BMPs and technology innovations in industrial water management. The three focus areas of the workshop were water efficiency, water treatment and reuse practices, and BMPs in water management. The agenda included presentations by the U.S. Environmental Protection Agency, Sandia National Laboratory, Siemens Water, The Coca-Cola Corporation, Texas Instruments, EPRI, the WateReuse Foundation, and the National Council for Air and Stream Improvement (US forest industry). The 45 participants represented industry, government, academia, water associations and foundations. Workshop participants identified 10 Top Priority areas for ASME to pursue. **Download a copy of the full report - *Water Management Technology Best Management Practices for the Process Industries*. Download a copy of the Executive Summary Only**.

Appendices - Water Management Technology Best Practices and Innovations for the Process Industries Report

Industrial Water Use and Reuse Workshop

April 30 - May 2, 2013

Grand Hyatt, San Antonio, Texas

Notes from Implementation Session

Loraine Huchler, Chair

Panelists

Role	Name
Facilitator	Loraine Huchler President MarTech Systems, Inc. huchler@martechsystems.com
Regulatory	Tara Lanier Marathon Petroleum tllanier@marathonpetroleum.com
Recycled Water Supplier	Joe Walters Manager of Business Development and Regulatory Affairs West Basin Municipal Water District joew@westbasin.org
Plant – Project Engineer or Utilities Engineer (also functions as the Capital Project Committee Member)	Don Akers Technologist Marathon Petroleum dhakers@marathonpetroleum.com
Consultant (Water Treatment Supplier in "real life")	Sid Dunn Baker Petrolite sidney.dunn@bakerpetrolite.com
Consultant	Jerry Levine Principal Water & Power Consult jerry08536@gmail.com
Equipment Supplier	Al Pristera VP Project Development, Water & Waste Solutions Ecolab Al.pristera@ecolab.com
Water Treatment Supplier	Tim Eggert Senior Technical Consultant, Regional Engineering, Water & Process Technologies GE POWER & Water Timothy.Eggert@ge.com
Water Treatment Supplier	Michael Goldblatt Applications Project Manager Ashland Chemical Company MGoldblatt@ashland.com

Two scenarios

- **Recycled Water at the Inlet to the Plant**
  - Title 22 water is recycled water that conforms to the water quality specification in CA statute "Title 22"
  - There are other sources of municipal or private recycled water, especially in areas of water scarcity such as the Middle East or Caribbean Islands
  - This scenario also applies to situations where the influent water to a plant has a significant and long-term or permanent change in the concentration of contaminants.

- **Internal Reuse**
  - Individual streams are recycled within the boundary limits of a plant
  - This is the more common scenario due to the absence of municipal or private recycled water suppliers in the U.S.

Roles

Panelist preparation: *think about your role and responsibilities as a member of a water reuse team. Think about your experiences in these projects and perhaps comment on a key insight or approach that you could offer as advice on tackling projects like these.*

- **Regulatory (Joe Walters, Tara Lanier):** Role is to ensure conformance of operations to Federal and State regulations regarding air and water discharges and be proactive about possible changes in permit requirements.

- **Project Engineer (Don Akers)**: Role is to
  - gather cost data to justify any capital expenditures
  - assess operating expenditures
  - meet the ROI hurdle
  - assemble the proper support team
    - finance
    - maintenance
    - operations (production)
    - engineering
    - environmental
    - sustainability
    - water treatment supplier (advisor)
    - corporate water treatment staff (if available)
  - consider additional stakeholders
    - Finance
    - Maintenance
    - Sustainability
    - Public
    - Public Relations
    - Identify the decision-maker
    - Company shareholders
    - Regulatory agencies
    - Health & safety

        - Water purveyor
        - Academicians, graduate students, undergraduates
        - Local politicians
        - Legislators
        - Environmental advocacy groups
        - Legal
        - "Mother Nature" – e. g. Climate
    - ensure that the project follow their corporate practices for project management, methods to analyze return on investment, criteria of success
    - identify and manage the project tasks, scope and budget
    - create a Request for Proposal (RFP) for Engineering Firm
    - establish negotiable items for design phase
    - participate in the acceptance phase during commissioning

- **Consultant (Jerry Levine, Sid Dunn)**: A consultant's role varies:
    - advisor: technology, applications, chemistry requirements, equipment selection
    - conduct feasibility study
    - ensuring proper integration with existing infrastructure,
    - ensuring operability
    - determining operator training requirements
    - assessing risks for the proposed system configurations
    - establishing best practices
    - contingency planning
    - assess impact on the wastewater treatment plant and discharge permit
    - liaise with regulatory agencies
    - validate costs and return on investment
    - substituting for in-house staff

- **Equipment supplier (Al Pristera, Tim Eggert)**: Role is to
    - identify the options for technology and
    - provide cost estimates and
    - delivery schedules for capital equipment.
    - pilot testing to confirm design and performance

- **Water Treatment supplier (Tim Eggert, Mikel Goldblatt, Sid Dunn)**: Role is to
    - assess impacts from the influent to discharge
    - identify chemical treatment programs to control deposition, corrosion and microbiological growth with water that has new or higher concentrations of contaminants
    - product trials are often required to ensure a good fit

Role Play

- **Justification of a project** (Joe Walters, Tara Lanier, Don Akers, Tim Eggert)
  Panelist preparation: *your manager has just told you that the plant is going to embark on a water reuse project, and indicated that your first responsibility is to identify a justification in preparation to sell the project to the project budget review committee. Explain how various drivers might be compelling (or not). You might want to draw on past experience (e. g. the multi-year drought in southeast Texas two years ago) or describe specific corporate cultures or industry trends.*

  - Economics
    - Lower cost water is available (e. g. Title 22)
    - Change (usually increase) in chemical treatment and operating costs
    - Change in risks (that always translate into costs – due to mitigation of consequence – either after they happen or avoidance measures)
    - Synergy with plant expansions that require additional influent or wastewater treatment assets
  - "Acts of God"
    - Local drought conditions cause poor water quality and a <u>temporary</u> reduction in withdrawal quantities.
  - Corporate Policy
    - Corporate mandate to reduce water usage within plant
  - Regulatory
    - Changes in regulatory issues: withdrawal volumes, discharge volumes, discharge contaminant concentrations
  - What do you do about intangible drivers
    - Sustainability
    - "good corporate citizen"
    - public-private pilot projects

- **Project Definition** (Don Akers, Tim Eggert, Jerry Levine, Sid Dunn, Al Pristera, Mikel Goldblatt)
  Panelist preparation: *consider that you must now "do your job" – analyze the problem, identify the required expertise, gather the data and determine the best opportunities for water reuse. Walk through this process with the team members – the other panelists – and describe the major obstacles and options as well the deliverable. Answer the question, "How do you provide the required information to the engineering design firm who will actually design the project?"*

  - Roles and Responsibilities
    - Local and corporate plant personnel
    - Outsource resources: consultants, equipment suppliers, water treatment suppliers, others?
  - Information gathering
    - Auditing existing water systems
    - Tools to map water balance
    - Creating a design basis

- **Hurdles**
  - Water reuse projects may be a hard sell to CEO's, especially in times of water abundance.
  - Proactive efforts to conduct water conservation are trumped by events like droughts.
  - The cost of fresh water, and even potable water, for both industrial and residential users, does not reflect the true cost of this scarce resource and does not attempt to modify the users' behavior through a supply-based pricing model.
  - The lack of a validated flow and "salt" balance for a plant increase the project cost and dramatically raises risk of failure for a water reuse project.
  - Failure to create a defensible design basis – a tabulation of the worst case, best case and average case for the influent water to a process – increases the project cost and dramatically raises the risk of failure for a water reuse project.
  - There are severe political pressures that oppose raising the price of water for constituencies that may not re-elect those commissioners.
  - Increased sludge volumes and increased waste disposal costs are sometimes an unintended consequence of water reuse projects.
  - The quality of treated wastewater is usually more dynamically variable than the quality of fresh water. In other words, the quality of fresh water varies seasonally with some predictability, while the quality of treated wastewater varies with different feedstock sources (e. g. crude) and with different efficiencies of the wastewater unit operations (especially the biological processes).

- **Drivers**
  - **Regulatory**
    - Lower discharge volumes
    - Stricter limits on maximum concentrations and total pounds per day of contaminants
    - Key contaminants: TSS, phenols, sulfides
    - Expect tighter limits on: nutrients, especially soluble phosphates, selenium, mercury
    - Regulatory Agency may not be willing or able to provide helpful guidance for plant owner regarding trends in permit restrictions or alternatives to resolve constraints to water reuse projects
    - NPDES permit writers need to be educated about the complexities of water reuse – lower volumes and, consequently, higher concentrations of contaminants. One issue is that the higher concentrations of contaminants may pose a threat to the ecology of the receiving watershed.
  - **Statewide Goals (Policy)**
    - Example: Southern CA
    - Several qualities of recycled water (treated wastewater) from the water authority
      - Title 22 – tertiary-treated disinfected water
      - Nitrified water (Title 22 without ammonia)
      - 1st pass RO (suitable for low pressure BFW)
      - 2nd pass RO (suitable for low and moderate-pressure BFW)
    - Water authority subsidizes the cost of the recycled water – calculate the cost to make it competitive as compared to fresh treated water, and consider the increased cost for the water in the circuit (increased corrosion and deposit control costs in various unit operations)
    - There are significant political hurdles to influence the economics by increasing the cost of fresh treated water to make recycled water more competitively priced and drive users to choose recycled water.

- **Opportunities**
  - Most water reuse projects are a part of a larger plant expansion – which means that there are several opportunities to justify the project. Retreating water within a unit:
    - may avoid an expansion of the pretreatment and/or wastewater treatment plant
    - may result in greater reliability and lower risk due to a more consistent and/or higher quality of retreated water
    - may result in energy recovery
    - may result in an economically neutral trade-off of cost of retreatment versus installation of transfer piping over long distances
    - the amount of capital money required to fund the "water reuse" portion of the project is small compared to the total capital cost of the project – perhaps allowing other justifications such as "responsible citizen" or other sustainability-related value to complement the economic justification (return on investment)
- **The value of scenario planning (a. k. a. contingency planning)**

  Scenario or contingency planning is often conducted for short-term catastrophic events, such as hurricanes, but seldom conducted for long-term gradual events like drought or climate change. In this context, the concept of scenario planning entails speculating about future events that would have negative impacts on the safety, reliability, operability and integrity of assets. Planners assign a relative score to each possible event based on the severity of the outcome (the "consequences") and the probability of occurrence (the "frequency").

  Lesson learned: The raw water reservoir at Valero's Port Arthur, TX refinery was located outside of the sea wall, making it vulnerable to inundation by seawater during a severe hurricane. Refinery staff conducted scenario planning and determined that this risk of seawater inundation and complete interruption of the fresh water supply for the refinery was extremely small and had a very low probability of occurring – a "low frequency, high consequences" event. Therefore, the refinery did not invest in any contingency efforts.

  The storm surge from Hurricane Ike inundated the raw water reservoir, preventing the refinery from restarting after the storm ended. Plant personnel estimated that draining and flushing the pond would require several months due to the large volume and the complexity of the task. A suggestion to use readily-available irrigation piping and special permission to withdraw water from a nearby canal provided approximately 20% of the normal raw water volumes and allowed the refinery to "warm up" the refinery using a single boiler within three weeks. Simultaneously, refinery personnel improved on the transfer piping idea to create a larger-diameter floating pipeline to provide sufficient volumes to restore 100% of the required raw water volumes. Within six weeks, the refinery was at full water flowrate using a temporary floating pipeline. After complete recovery from the hurricane, refinery personnel "hardened" the pipeline to make it a permanent contingency.

  This case history illustrates that a crisis management event can sometimes result in a capability that normal contingency planning could not justify using traditional risk assessment metrics.

- **Execution of a project** (Don Akers, Tim Eggert, Jerry Levine, Sid Dunn, Al Pristera, Mikel Goldblatt)
  Panelist preparation: *think about the design phase. How did you select the engineering design firm? What are the roles and responsibilities of team members? How much oversight and input do you want or need regarding the engineering design (see section "Engineering design" below)? What about commissioning complexities? How will you validate performance of the system, especially if the water quality differs from the design basis?*

    - Business decisions
        - Capital purchase, lease, insourced or outsourced operation
    - Engineering design
        - Selection of engineering design firm
        - Key resources needed
        - Technology selection
        - Operability analysis
        - Maintainability analysis
        - Integration issues
        - Identification of negotiable and non-negotiable features
    - Commissioning
        - Shutdown or rental equipment to prevent lost production
        - Key resources needed
        - Validation of performance

- **Project Management**
    - Process safety management has made the engineering and management of projects much more rigorous.
    - The organization of teams, communications, workflow and even the terminology may vary based on the culture of the project owner's organization.
    - Identifying the stakeholders, defining the design basis, the hydraulic and "salt" balance and the scope of work and explaining the outcomes and criteria for success are the first tasks for a project. Failure in any of these tasks will result in increased costs and possible delays – those dreaded "change orders."
    - As the project team defines the scope of the project, it's critical to consider the near-term plans for expansion, process reconfigurations or idling of units within the plant. These projects typically require four or more years from initial conception to implementation.
    - Once the budget committee authorizes the project, the next step is to identify and ensure engagement of the two key owner roles in a project: the project manager and the technical lead.
        - The project manager is the person whose mandate is to be under budget and on schedule.
        - The project lead is the "technical decision maker," responsible for ensuring that critical design elements and system capabilities aren't negotiated away as the project seeks to stay on budget and on schedule.
    - The engagement of the project lead and other stakeholders such as the water treatment supplier and a consultant, if desired, are critical as the team prepares the Request for Proposal (RFP) and qualifies the engineering design firm. These professionals will collaborate to ensure the accuracy of the information included in the RFP and the accuracy and operability of the proposed design. Integration issues are often a critical issue for operability.

AUTHOR INDEX